AN AIR FORCE HISTORY OF SPACE ACTIVITIES

1945 - 1959

By
Lee Bowen

August 1964
USAF Historical Division Liaison Office

SHO-C-64/50

When this study is no longer needed, please return it to the
USAF Historical Division Liaison Office, Headquarters USAF.

FOREWORD

Covering the efforts of the United States from 1945 to September 1959 to wrestle with the unknown ramifications of space, this history includes both the civilian and military activities. An Air Force History of Space Activities presents a more detailed treatment of information published in 1960 under the title of Threshold of Space, 1945-1959. Other monographs on this subject include The Air Force in Space, 1959-1960, and (in draft) a sequel for fiscal year 1961.

The author of this history begins with the work of the early pioneers in rocketry, the first satellite feasibility studies by the military, and the relationship of the ballistic missile to the space vehicle. He reviews the Russian and U.S. space programs between 1945 and 1957, during which efforts were made to create space law and the United States chose to pursue a space-for-peace policy. The conservatism of policy makers raised obstacles, but there were space projects, some of them under the Air Force.

After the shock of Sputnik I, the reshaping of policy resulted in the establishment of ARPA in the Department of Defense and NASA as the civilian space agency. The author tells of ARPA's supremacy over the military services in 1958; its loss of control to NASA in October 1958; NASA's activities from then until July 1959; the position of the Air Force after losing out to both ARPA and NASA; and the Air Force's determination to cooperate with NASA, through research, development, and use of its facilities. Within the DOD in 1959, authority for space research and development was transferred from ARPA to DDR&E, interservice tension mounted, the Air Force struggled to regain lost projects and objected to Navy's appeal for a military space command, and the tide turned for the Air Force when the Secretary of Defense decided in September to give to it the responsibility for the development and launching of all DOD space boosters and for management of Sentry, Midas, and Discoverer.

JOSEPH W. ANGELL, JR.
Chief, USAF Historical Division
   Liaison Office

CONTENTS

I. The State of Rocketry, 1945-57... 1
    The Emergence of Rocketry as a Science........................ 1
        Pioneer Thinkers.. 2
        The Great Work of Peenemuende............................ 3
    The Military Missile Program.................................. 7
        The Postwar Attitude toward Research and Development..... 9
        Rocket-Engine Research................................... 13

II. The Military Services Plan for Space, 1945-57...................... 25
    The Russian Space Program..................................... 25
    Service Moves toward Space Projects and Policy, 1945-48....... 29
    Resume of 1947-57 Rand Studies................................ 39
    Air Force Space Projects, 1948-57............................. 43
        The X-15 Research Aircraft............................... 44
        Bomi--Robo--Brass Bell--Hywards--Dyna Soar............... 45
        The Advanced Reconnaissance System (ARS) or WS-117L...... 46
        Hypersonic Environment Test System....................... 48
        Man-in-Space (MIS)....................................... 50
        Propulsion... 54
    The First Army-Navy Space Project............................. 55

III. A Policy of Space-for-Peace--and its Effects...................... 57
    The First Concepts of Space Law............................... 58
    First Efforts to Create a Space Law, 1955-57.................. 61

IV. Lost Opportunities--Vanguard and Others........................... 68
    The President's Decision to Support a Space Project........... 69
    Specifications for Vanguard................................... 71
    The Unnecessary Delay... 75
    Last Possibilities for a U.S. "First"......................... 77

V. A. Post-Sputnik Shaping of Policy.................................. 82
    At First, Emotional Reactions................................. 83
    A Modified Space-for-Peace Policy............................. 86
    A Compromise Space Program at Home............................ 93

VI. Organizing for the Military Space Program......................... 99
    Advanced Research Projects Agency............................. 102
    Director of Defense Research and Engineering.................. 107
    Activation of Directorate of Advanced Technology.............. 109

VII. Civilian Space Agencies.. 113
    Hearings on the President's Proposed Space Agency............. 114
    The National Aeronautics and Space Act of 1958................ 116
    Organizing Space Agencies under PL 85-568..................... 119
    Organization for Space at End of 1958......................... 122

VIII.	Eight Months of ARPA Supremacy..	124
	The Sources of ARPA's Program.....................................	124
	Vanguard and Explorer I......................................	125

VIII. Eight Months of ARPA Supremacy.................................... 124
    The Sources of ARPA's Program............................. 124
        Vanguard and Explorer I................................. 125
        Air Force Requests for DOD Approval..................... 127
        Recommendations of the Advisory Group on Special
            Capabilities... 127
        The USAF Astronautical Program of 24 January 1958....... 131
        Last USAF Efforts to Save the Astronautical Program..... 133
    The Rule and Program of ARPA in 1958....................... 136
        ARPA's Operating Procedures............................. 136
        The Assignment of Projects to ARPA...................... 138
        ARPA's Assignment of Projects........................... 141

IX. NASA's First Program, October 1958 to July 1959................ 145
    ARPA's Claim to Border Projects............................ 146
    The President's Division of Space Projects................. 148
    NASA's First Nine Months................................... 149
    NASA-USAF Relations.. 155

X. ARPA and the Military Program, October 1958 to July 1959....... 162
    The Military Space Program--Second Phase................... 162
        The Satellite Projects.................................. 164
        The Booster Program..................................... 168
        The Tracking System..................................... 170
        At the End of June 1959................................. 174
    ARPA's Changing Status in DOD.............................. 175
    Space Operations, October 1957 to July 1959................ 177

XI. Air Force Space Policy and Supporting Action, 1957-59......... 181
    Military Reactions to the Space-for-Peace Policy........... 183
    The Doctrine of Aerospace.................................. 187
    The Navy-Air Force PMR Disagreement........................ 191
    A Time for Decision.. 196
        An Air Force Attempt to Force the Issue................. 197
        The Navy's Appeal for a Space Command................... 198
        ARPA's Move for a Mercury Joint Task Force.............. 199
        McElroy's Decisions of 18 September 1959................ 200

NOTES.. 205

GLOSSARY... 226

TABLES

Testimony of Neil H. McElroy, SOD, 27 November 1957........... 104

The Air Force Astronautical Program of 24 January 1958........ 132

Air Force Fund Status Summary, February 1958.................. 135

NASA's Program for FY's 1959-60............................... 151

ARPA's Program.. 163

I. THE STATE OF ROCKETRY, 1945-57

Despite astronautical fancies during the centuries of mythology, and even after the coming of respectable astronomy, it was not until the latter part of the nineteenth century that there was any understanding of how to launch a man-made object into space. Jules Verne was the first writer to put fictional space travel on something of a rational basis. In his book <u>From the Earth to Moon</u> he employed a piece of specially designed artillery to project a manned shell within grasp of lunar gravity. His approach was fundamentally correct--the use of a ballistic missile--but his "powerplant" imposed insuperable difficulties and provided no means for a "soft landing."

It is surprising that a man of Verne's imagination did not think of using rockets because their behavior had been observed for generations. Rockets had even been turned to military ends by Sir William Congreve, and they played a spectacular, though ineffectual, role in the Napoleonic wars and in the British-American war of 1812. Naturally the military showed no enthusiasm for a weapon that did little more than frighten inexperienced troops. There followed a century in which the military were generally skeptical of any practical use of rockets.

The Emergence of Rocketry as a Science

From 1815 to 1900, rocketry was largely discarded except as a night ornament of fireworks for national holidays. Then, between 1900 and 1930, a number of reputable scientists and engineers voiced a new faith in rocketry as a source of propulsion for missiles, aircraft, and even spacecraft. Two characteristics made the rocket the necessary powerplant for flight above the atmosphere. First, along with other reaction engines, it did not depend upon friction for the movement of its vehicle. It depended solely upon the principle originally described by Sir Isaac Newton in his third law of motion: for every action there is an opposite and equal reaction. Second, the rocket engine, unlike the other reaction engines, was not air breathing. It breathed the oxygen of its own fuel. So long as the fuel was not exhausted, the engine could function in the near vacuum of space quite as well as within the atmosphere.

Pioneer Thinkers

The significance of rocketry was not understood until the pioneer thinkers of the early twentieth century began to see the implications. Among these thinkers, three were outstanding because they could express their ideas in precise mathematical formulas: Konstantin Eduardovitch Ziolkovsky, Robert Hutchings Goddard, and Herman Oberth.

Ziolkovsky, a reticent, self-taught Russian scholar, was the first to perceive the significance of the rocket as the only engine that would permit a deep penetration of space. In 1903, he published in the Russian periodical Science Survey an article on rocket propulsion for space vehicles. He supported his thesis with remarkable calculations in mathematics based upon a thorough knowledge of astronomy, physics, and chemistry as those sciences were then known. His work excited some interest within a small, almost esoteric group of compatriots. The publication of other learned articles during the next 10 years earned him increased respect among Russian scientists, but his fame did not go far beyond his own country. Few western Europeans knew Russian in the early 1900's, and translations of Ziolkovsky's works were not impressive.

In 1919, Goddard, an American, prepared his now famous report but one largely disregarded at the time--"A Method of Reaching Extreme Altitudes." He too, like Ziolkovsky, supported his thesis mathematically, and he went the Russian one better by firing a revolver in a vacuum to test the recoil and prove experimentally that Newton's third law of motion would be applicable to objects in space. Though he devoted most of his life to the promotion of rocketry, he won few disciples. He remained almost unknown, both at home and abroad, and took his place in the long line of prophets without honor.

In Germany, Oberth fared somewhat better. In 1923, uninfluenced by Goddard, he published his own scientifically developed thesis on rocketry--The Rocket into Interplanetary Space. He included designs for rocket-propelled vehicles and advocated the use of liquid fuels as superior to the dry propellants previously employed. Almost at once there emerged numerous enthusiasts at home and abroad who, with slight grasp of the subject, took up the somewhat premature hobby of space travel. In Europe and America, space societies sprang into being, peopled by visionaries who mistook themselves for space literati.

The Great Work of Peenemuende

The German army, in the late 1920's, still trying to break its Versailles straight jacket, was more realistic in its approach when it became interested in rockets as possible weapons. In the dearth of practical and scientific knowledge, the soldiers of the Second Reich, though possessing very limited facilities, determined to develop and prove the worth of rocket engines.

In 1930, German military authorities selected Capt. Walter Dornberger, a technologist of marked ability, to develop in utmost secrecy a liquid-fuel rocket with a range exceeding that of existing artillery. Dornberger, directly responsible to the chief of the German army, began his work at a small proving ground about 20 miles from Berlin. His organization consisted of a small staff of officers who directed a much larger number of civilians. Six years later, Dornberger moved his group to Peenemuende on the Baltic Sea where he established the Rocket Experimental Station, dedicated to the development of radically new weapons that, if successful, would give Germany dominance in Europe.[1]

The near success of Peenemuende is an impressive tribute to the competence of Dornberger and his staff. The experimental station grew rapidly. At its peak it had 18,000 employees, and its area covered 20 square miles. The work was all-inclusive from basic reasearch through development, production, training of specialized troops, and eventually supplying the front formations with finished weapons. Dornberger and many of his associates seem always to have been aware of the full potentiality of rockets. Thinking of developmental stages rather than of military or scientific ends, the Peenemuende staff envisioned a program that would move systematically from short-range missiles to an eventual goal of space vehicles. Ultimate success depended upon the realization of adequate powerplants and looked into the future for nuclear, ion, and proton propellants.

Dornberger's support was not lavish but sufficient to permit swift progress. In 1936, he reported the feasibility of liquid-fuel engines, and six year later, 3 October 1942, he conducted the first successful launching of the A-4 missile, now generally referred to as the V-2. It demonstrated a velocity of about 3,600 mph and an altitude of 55 miles, far beyond the reach of air-breathing engines. The V-2 was a long way from the space vehicle or the satellite foreseen by Dornberger, but with proper priority the missile could have supplied the Nazi government with a possible war-winning weapon.[2] The V-2 might also have become a direct threat to the United States, for Eugen Sänger, a Nazi-employed scientist, suggested that it be used as second stage of a boost-glide vehicle to be launched in Germany against New York City.

Fortunately for the Allies, Hitler was long indifferent to the accomplishments of Peenemuende. It was not until the experience of Stalingrad shook his self-confidence that he turned to "secret weapons" to save him. On 13 June 1944, a week after D-day for Operation Overlord, the Germans fired the first of the V-1 guided missiles against England. The weapons were disturbing, but they were much less so than the V-2 rocket-propelled ballistic missiles that began a bombardment of France, Belgium, and England on 8 September. By that time the Allied armies were well established on the continent and within weeks would capture the more strategically located missile bases, thereby reducing the enemy threat. But Gen. Dwight D. Eisenhower knew that six months earlier the V-1 and especially the V-2 might have endangered the Anglo-American invasion of France.[3]

The collapse of Peenemuende began soon after the launching of the V-2 offensive. By autumn 1944 the war was sweeping toward the Nazi defeat. The Russians raced through eastern Germany and in January 1945 threatened Peenemuende. In the confusion of disaster, Dornberger could not save his entire group, and a number of his employees, some with 10 years' experience,

had no choice but to remain at the experimental station. A small section
that fortunately included Dornberger and some of his most highly qualified
specialists escaped westward to the Hartz Mountains. They took with them
invaluable papers and established a new installation that functioned only
in the sense of holding together scientists who might otherwise have been
scattered and lost.[4]

In the spring of 1945 the Russians occupied Peenemuende. They transported to Russia 4,000 scientists and workers along with most of the
equipment, production facilities, and as many documents as they could find.
Inside the Soviet Union the Russians reconstituted Peenemuende in a diminished version and operated it as a rocket research center. The captured
Germans manufactured new models of the V-2 and in this way between 1946
and 1948 supplied the Russians with missiles for at least 500 experimental
firings. By that time, the Soviets felt they had drained the Germans of
all possible contributory knowledge and, after a long period of isolation,
permitted their prisoners to return home. The Germans remained ignorant
of the Russian competence in the field of missile propulsion and knew
nothing of the Russian overall rocket program.[5]

When the Americans overran the Hartz Mountains in May 1945, they
found quantities of the Peenemuende records and, even more important,
some of the men responsible for the work. At once the U.S. Army seized
the documents, blueprints, and data and shipped them to the United States.
Soon afterwards the Army initiated Operation Paperclip and brought to the
United States 180 of the scientists who had played leading roles at
Peenemuende.

Having thus scavenged the broken body of Peenemuende, the Russians and Americans reacted very differently to the taste of their spoils. The Soviet government understood that rockets would be of paramount importance in the future and directed their nascent program toward nothing more definite, and nothing less inclusive, than the advancement of rocket science regardless of its specific applicability. The result was that by 1956 or 1957 the Russians had a rocket engine--or possibly rocket engines--with a thrust that could launch either missiles laden with atomic warheads or heavily instrumented capsules to orbit the earth or explore interplanetary space. The Americans, in contrast, had little top-level guidance or support and they fragmented their development of rocket engines among a number of projects within a comprehensive but frequently unstable missile program. The unfortunate consequence was that in 1956 and 1957 the United States had no rocket propulsion comparable to that enjoyed by the Soviet Union.

The Military Missile Program

A missile is "any object thrown, dropped, projected, or propelled, or designed to be thrown, dropped, projected, or propelled for the purpose of making it strike a target."[6] According to this definition, missilry is a genus of many subordinate species. Insofar as the word is used attributively in the expression "guided missile program"--often shortened to "missile program"--it denotes the development of self-propelled, unmanned vehicles, armed with warheads to be sent against enemy targets and equipped for guidance either by a preset device or radio command.

To shortcut through the semantics of listing "ballistic missiles" under the species of "guided missiles"--a confusing contradiction in terms--it should be sufficient to say that the postwar "missile program" came to include any wholly or partially guided missile using a turbojet, ramjet, or rocket powerplant or any combination of them. Whether the missiles were aerodynamic in structure was unimportant as far as the space program was concerned. The important fact was that only one of the three forms of missile propulsion could function in areas above the operational altitude of air-breathing engines. Therefore, those missiles designed for trajectories passing through and continuing above the atmosphere had to be propelled by rocket engines. These rocket-propelled vehicles were generally referred to as "ballistic."

A quirk of history brought together the ballistic missile program on the one hand and the satellite-space program on the other. In 1945, when the armed forces, along with the rest of the world, tried to adjust to the new era of technology, there was still little thought, except in very limited circles, of a near-future breakthrough into space. Certainly in the spring and summer, when the remnants of Peenemuende were being absorbed by the Russians and Americans, there was no American space program. In the early part of 1946, both the Navy and the Air Force came forward with satellite feasibility studies, but neither proposal reached the stage of research and development; and in the years between 1946 and 1954 nothing was done to design and produce a rocket engine for satellite purposes. In 1954 and 1955 the space age had assumed an imminence totally lacking 10 years before. Since there was still no specially designed satellite engine, the services turned realistically to what was or soon would be available in the military missile program as it had developed since 1945.

Of course only the development of rocket engines is pertinent, and that lies always within the more extensive program of "guided missiles." The problem then is to ferret out from a labyrinth of complexities the decisions that delayed or hastened the research, development, and production of "ballistic missiles" and to relate these decisions to technological difficulties and national policies.

The Postwar Attitude toward Research and Development

During World War II the Army Air Forces (AAF) and the Army Service Forces (ASF), particularly the Ordnance Department, each sponsored a missile program. The accomplishments were far less important than those of Peenemuende. Nevertheless both AAF and ASF went far enough in the development of research and weapon carrying missiles to provide invaluable experience in meeting the early postwar demands that came as a consequence of the demonstrated effectiveness of the V-1 and V-2. The postwar program started off well but soon ran into technological and funding difficulties. Poverty was especially persistent and was without remedy for almost a decade. Indeed, lack of funds was one of the major reasons for the retarded development of rocket engines.

The American shift from a wartime to a peacetime way of life was translated into an economy program for the military that amounted to austerity in some areas and to retrenchment in almost all others. In April 1958, in the midst of American chagrin over the Sputnik incident, Dr. Wernher von Braun, Director of the Development Operations Division, Army Ballistic Missile Agency (ABMA), attributed Russian success to the Soviets' continued postwar emphasis on military needs, whereas the United

States chose the production of consumer goods.[7]

Retrenchment is particularly damaging when applied to basic research, where neglect, or even inconsistency of effort, can result in time losses for which crash programs can offer no more than partial compensation at best.[8]

The defense research program ran into trouble as soon as World War II ended. The armed forces were disturbed--they understood what the cutback might lead to, but there was little they could do about it. An improvement followed the outbreak of the Korean War but did not survive the conflict. After January 1953, military research and development and the position of scientists in government employ suffered further deterioration.

Meanwhile, Senator Joseph McCarthy, of Wisconsin, and his like were in the heyday of their rampancy. Their vicious anti-intellectualism coincided, in time only, with the more kindly jibes of Secretary of Defense Charles E. Wilson, bent upon defending the administration's policy of economy and downgrading of research and development. Accepting the sufficiency of existing weapons, Wilson could see small justification for a basic research that might contribute to unforeseeable requirements in the future. The Secretary became a tireless spokesman against "boondoggling research."[9] One example of his frequently expressed opinions will show his point of view:[10]

> In my own mind I think of . . . /research and development/ like drilling for oil . . . . The smart people in the oil business try to drill their holes for oil in a likely place, so the money that is given to the Defense Department for research and development, I like to see spent in an area /where/ if . . . successful . . . it will be of some use to us. And maybe some other place in the nation's budget could go the money for fundamental research, I don't know. I don't care what happens to some of the minor other things.

The effect of McCarthy persecutions and the administration's skeptical attitude toward research and development persuaded some very able and competent scientists to withdraw voluntarily from government service. The harm done to the United States cannot be estimated but it was not negligible. There were protests at the time, but the voices of dissent--some of them highly placed--were disregarded. Senator W. Stuart Symington; James H. Douglas, then Under Secretary of the Air Force; Gen. Nathan F. Twining, Chief of Staff, USAF; Gen. Thomas D. White, Vice Chief of Staff, USAF; and Lt. Gen. Donald L. Putt, Deputy Chief of Staff/Development, were among the notables who warned that the United States could not match Russian military superiority in numbers. National security depended upon maintaining a superior technology, which, in turn, required extensive and unceasing research and development. The Russians for their part were concentrating heavily upon a progressive program of research and development, and their success seemed phenomenal. At least twice, in 1953 and again in 1956, National Intelligence Estimates stated that the Soviets would have ICBM's by 1960 and that they already had some missiles capable of carrying nuclear warheads.[11]

The warnings were repeated at yet higher levels. On 24 September 1954 and again on 8 September 1955 the National Security Council (NSC) declared that "basic and applied research must be kept abreast of the changing Soviet threat" and that Soviet possession of ICBM's before the United States produced them "would have the gravest repercussions on the security and cohesion of the free world." Paradoxically, the President approved both NSC declarations, but the administration's policy remained

---
* Among the outstanding American scientists who suffered from McCartheyistic aspersions were Dr. Harold Urey, Dr. Edward U. Condon, and Dr. Allen V. Astin.

unchanged.[12]

The climax of protests came late in 1955 and early in 1956. Trevor Gardner, Assistant Secretary of the Air Force (R&D), spoke out strongly against inadequate budgets for research and development. He called for the fiscal year 1957 outlay to be doubled to meet the challenge of Soviet technology. His warnings ignored, Gardner resigned to show the intensity of his conviction.[13]

To this criticism, Wilson invariably replied in the same tone. Undoubtedly the Russians did have atomic bombs, 200 divisions, and numerous submarines: "So what? The final thing is, is there any reason for them to go to war and if they did, would it be clear to them that they would meet so much opposition that they would really lose." Wilson based his optimism on the Air Force, which he described as "the best in the world." It would be tragic, he said, if the Air Force were second best. The Secretary saw the Air Force equipped with currently available weapons; he showed little sign of seeing that without a strong and unending program in basic research those weapons might soon become obsolete and the Air Force weak.[14]

On 8 February 1956, almost exactly five months after approving NSC's latest call for an expanded research program, Eisenhower expressed astonishment at the concern of Americans because of Russian success. He was sure that the United States was ahead of the Soviet Union in all important aspects of "guided missiles."[15] Wilson thereupon introduced a new if uncomprehensible theme. His argument ran that since missiles were phychologically as well as militarily significant, they should not be overemphasized. Shortly afterwards, Donald A. Quarles, Secretary of the Air

Force, said that the U.S. investments in military research and development were more than adequate.*

Rocket-Engine Research

Between 1945 and 1955 the Army and the Air Force continued the wartime goal of developing both research and weapon-carrying missiles. The effect of V-2 in both fields of endeavor was to give an interest in rocket propulsion that previously had too often been totally absent. However, neither the high-altitude rocket research nor the development of engines for military missiles sought the ultimate in rocket significance. Each program followed the practice of developing rocket engines to meet specific requirements.

Of course in the high-altitude rocket research program there was no need for an engine with a very large thrust. At most, a research rocket was expected to propel a payload of perhaps 1,000 pounds to an altitude of 200 or 250 miles. In the program for military missiles, planning did not exceed the concept of adjusting World War II experience to more

---

*The statistics for research and development are interesting. For fiscal years 1953, 1954, and 1955, the total appropriations for DOD research and development were, respectively, $1.6 billion, $1.4 billion, and $1.3 billion. The difference of $300 million between 1953 and 1955 is not great in postwar reckoning, so there seems to be only a slight decrease in research and development spending. But the significance lies not in the loss of $300 million over a period of three years but that there was no increase. Percentage-wise, the RAF was making double the investment in research and development during these years. To make the American effort commensurate with that of the RAF, the 1953 appropriations of $1.6 billion should not have decreased to $1.3 billion in 1955 but should have increased to $2.2 billion.

efficient methods of weapon delivery. In 1945-46 the Army thought of missiles as an adjunct to artillery; the Air Force thought of missiles in the traditional airpower terms of altitude, range, and velocity. Because almost no target was more than 5,000 miles away and because the Atomic Energy Commission soon undertook to bring nuclear warheads well below the 10,000-pound weight of wartime bombs, practically no thought was given to the development of missiles with a greater range or a greater capacity. The greatest powerplants envisioned did not exceed the requirements of intercontinental strategic missiles; that these requirements far exceeded the range, altitude, velocity, and capacity of the V-2 is not the point. It was the limitation imposed upon rocket-engine development by the missile program that did no much, later on, to hold back the program for satellites and other space vehicles.

High-altitude rocket research was conducted for the most part at the Army's White Sands Proving Ground in New Mexico. The White Sands experiments, however, were not limited to the Army, and both the Navy and Air Force participated.

Out of the wartime missile effort there had come the WAC-Corporal with a maximum altitude of 70 miles and a payload capacity of 25 pounds. Its first flight occurred at White Sands on 26 September 1945. By that time a few of the captured V-2's were reaching the United States, and they were entrusted to the Army for assembly and firing. Because of their greater capacity, the German missiles led to a waning interest in WAC-Corporal, but they also stimulated a more ambitious Army-Navy-Air Force upper-atmosphere research program, formulated in the late days of 1945. Ordnance then contracted the General Electric Company to conduct the V-2 experiments, and on 16 April 1946, GE representatives, assisted by scientists from Peenemuende, fired a V-2 for the first time from American soil.[16]

Experience with the V-2 soon showed a number of unanticipated shortcomings in the missile. The airframe was clumsy and unnecessarily heavy because of internal structural support. The tanks used were already obsolete by American standards and were a serious drag on the missile's proficiency. Moreover, since the V-2 did not lend itself to instrumentation, useless ballast had frequently to be added to the payload with unpredictable effects on the missile's performance. An acceptable interim solution was found in Project Bumper--mating the V-2 with WAC-Corporal in a two stage missile--but the arrangement only emphasized the need for a more satisfactory replacement. Moreover, the supply of V-2's was limited and this fact too was instrumental in persuading research authorities to call for a newly designed missile at an early date. They wanted one approximately as large as the V-2 but amenable to instrumentation. In August 1946 the Naval Research Laboratory (NRL) contracted the Glenn L. Martin Company for an improved rocket, which was called Viking. Also in 1946, with the Navy as cognizant agent, Douglas Aircraft Company began development of the smaller Aerobee.

Both the Viking and Aerobee were operational by 1948, and both had rocket powerplants. Both missiles were also dynamic and underwent numerous changes to meet the standards of new knowledge. In addition, Aerobee, at the request of the Air Force, was produced in a different version by 1955, known as Aerobee-Hi. All three missiles--Viking, Aerobee, and Aerobee-Hi--were used to perform experiments for the Army, Navy, and Air Force.

The following table indicates the contributions made to the development of rocket engines by the upper-atmosphere research program between 1946 and 1956:*

Missile	Length	Diam	Takeoff Gross Weight	Max Alt in Miles	Thrust in Pounds	Veloc: mph
V-2	46'6"	5'5"	28,500	132	56,000	3,600
WAC-C	14'4"	12'	665	70	50,000	2,800
Bumper			28,918	244	106,000	5,150 (2d stage)
Viking	42'8"	3'9"	16,000	154	21,000	4,000
Aerobee	23'6"	15"	1,500	38	18,000	2,100
Aerobee-Hi	23'6"	15"	1,300	164	18,000	4,600

Military (weapon) missiles had a quite different history from those of the high-altitude research program, and of course their purpose was quite different. But here, too, rocket engines went through a steady development based on the V-2, and the progress made between 1946 and 1955-57 became a determining factor in the nature of space operations between 1954 and 1959.

In addition to the advent of the V-2, there was another World War II event of great influence in the post war development of missiles and

---

*During 1946-56 the services developed a number of other research missiles--Rockoon, Rockaire, Dan, Nike-Cajun, Terrapin, Asp, and Wasp. They reflected the changing techniques in rocket research and cost only a fraction of the Aerobee and Viking. They were valuable in furthering upper-atmosphere research but contributed little or nothing in advancing rocketry to where it could put a satellite in orbit.

indirectly of rocketry--the use of atomic weapons. Immediately after the bombing of Hiroshima and Nagasaki in August 1945, there was talk of a marriage between a nuclear warhead and a missile vehicle to create the "ultimate weapon." Gen. H.H. Arnold, Commanding General, AAF, expressed his conviction that the atomic bomb would be integrated with guided missiles, and the Joint Chiefs of Staff (JCS) stated that the atomic and missile programs should be developed in such manner that each made use of the other. This was brave talk in the glow of victory, but it carried little weight in the repetition of post-World War I efforts to "return to normalcy."

In the late days of 1945 and throughout 1946, AAF put on paper a comprehensive guided missile study program that included 28 projects. Three of the proposed missiles involved rocket propulsion--the Consolidated Vultee Aircraft (Convair) MX-774 that called for a rocket missile with long range; the North American Aviation MX-770 supersonic surface-to-surface, 500-mile-range, glide rocket missile; and the Republic Aircraft MX-773, 1,500-mile supersonic missile that could be either ramjet- or rocket-propelled.[17] Altogether, it was a program pregnant with important possibilities if carried through without serious interruptions. But serious interruptions were at hand.

In December 1946 the administration decreed a cutback in spending for defense research and development during fiscal year 1947, though the year was already half gone. The Air Force was then faced with the unhappy task of deciding where its own cuts should be made. The first decision was a reduction in the missile research budget from $29 million to $13 million. The next decision was to weed out the least promising missile projects and keep the remaining ones under the accepted ceilings.[18]

Since experience at White Sands was already showing the V-2 less satisfactory as a weapon carrier than had been believed and the Convair MX-774 studies indicated a long development period, enthusiasm for the rocket missile waned considerably. The requirement for fuel of high specific impulse not then available, guidance difficulties, and unsolved problems of reentry indicated "a long series of costly experiments" that would eat deeply into the curtailed missile research and development funds. On 6 May 1947, Maj. Gen. Benjamin W. Chidlaw, Air Materiel Command, wrote the commanding general of the AAF:[19]

> The Air Materiel Command and AAF missile contractors for the past six months have been carrying on detailed studies of the probable cost of developing guided missiles. From these studies, the conclusion must be drawn that the AAF program, while desirable and technically sound, is considerably overexpanded if we are to carry on with . . . /our presently reduced/ budget. . . . /The/ AAF program must be drastically cut. This is believed best accomplished by eliminating all the so-called "insurance" missiles such as subsonic missiles to perform the same mission as supersonic missiles being developed. Also eliminated is the 5,000-mile-range rocket which does not promise any tangible results in the next 8 to 10 years /and the unpromising 1,500-mile rocket or ramjet missile of Republic/.

The next month the Air Force canceled its contracts for Republic MX-773 and Convair MX-774.[20]

By July 1947 the most important project remaining in the Air Force missile program, as far as the later space program was concerned, was the MX-770, which soon came to be known as Navaho. Its development was subsequently conditioned by four major amendments to the contract between 1947 and 1950. In 1947 the plan was extended to cover three missiles-- the original 500-mile glide rocket missile plus a 1,500-mile missile and a 5,000-mile missile depending upon ramjets for cruising. In 1948 the Air Force abandoned the glide missile and rewrote the contract to provide for three ramjet missiles with ranges of 1,000 miles, 3,000 miles, and 5,000 miles. In 1949 the contract specified a 1,000-mile and a 1,700-mile ramjet air-launched missile and a 5,500-mile missile with a type of launching still undetermined. In 1950, at the suggestion of North American, the Navaho contract was changed again to provide for a three-step development beginning with a turbojet test vehicle, a 3,600-mile rocket-launched, ramjet-propelled experimental missile, and the ultimate 5,500-mile version.[21]

Despite the Air Force rejection of the rocket-powered glide missile in 1948 and the change to ramjet engines for cruising, North American continued the rocket development as a Navaho booster unit. The work became an important contribution to the accumulation of experience with rocket engines that, along with the accomplishments of the high-altitude research rockets, was indispensable to the later space projects.

In April 1946, when the MX-770 contract was signed, North American was without talent in rocket engines. To hasten the buildup of skill in this new field, the Air Force turned over to the contractor two V-2 rocket motors transported from Germany. They were excellent guinea pigs that North American engineers could study and from which prepare a Chinese copy brought up to the standards of American industry. From that beginning it should be possible to turn out a new, more efficient, and more powerful engine. Between 1946 and 1953, North American developed three rocket engines and designed a fourth, and a comparison of them with the

V-2 model reveals the progress being made in rocketry:[22]

Engine	Purpose	Thrust Pounds	Tested
German V-2	To propel a weapon missile	56,000	1946 at NA
XLR41-NA-1 (Chinese copy of the V-2)	Experimentation with a rocket engine brought up to the structural standards of American industry	56,000	1949
XLR43-NA-1	Designed to have a thrust of 75,000 pounds and intended to serve as the powerplant of the glide missile	81,000	1950
XLR43-NA-3	A smaller, lighter, and more efficient version of the XLR43-NA-1	130,000	1952
XLR71-NA-1	A combination of 2 XLR43-NA-3's to boost the Navaho 1,700-mile missile	240,000	1953
XLR71-NA-3	Designed as a booster for the 5,500-mile Navaho	400,000	1953 (in preliminary design only)

It is ironical that the success of North American in developing auxiliary rocket engines for various phases of Navaho should, along with rocket development elsewhere, have been one of the factors leading to the eventual cancellation of the Navaho project. By 1957, long-range, air-breathing missiles could be regarded only as standby weapons during the waiting period for operational ballistic missiles. Unfortunately for Navaho the Air Force had another long-range, air-breathing missile, Snark, which, though inferior in performance to Navaho, was considerably nearer operational status. It was pointless to continue the development of both Navaho and Snark. Under the circumstances the Air Force regretfully canceled Navaho, mindful of the great contributions North American had made to the science of missiles and rocket engines.

The 1946 policy of retrenchment had not dimmed the Army's interest in missiles, and especially rocket-propelled missiles, as adjuncts of artillery. Then, after 1949, when the Atomic Energy Commission made promising reports of lightweight atomic warheads in the near future, the Army initiated the Redstone ballistic missile, which, along with several other missiles of the Navy and Air Force, could be used to deliver the weapon. Redstone, however, had a range of 175 miles, and this seemed to be stretching the adjunct-to-artillery theory to cover a multitude of ambitions.[23]

The whole course of events--North American Aviation progress on rocket engines, AEC promises for smaller atomic warheads, some favorable Rand Corporation studies on the feasibility of ICBM's, the lengthening range of the Army's ballistic missiles--induced the Air Force to reconsider the advisability of a program for long-range rocket missiles. As early as 1951 the Air Force turned to Convair and requested a comparative study, to be completed within six months, on a rocket glide missile and a rocket ballistic missile. There was not much doubt, as far as Convair was concerned, what the recommendation would be. Since 1948, Convair--using its own funds--had continued small-scale work on MX-774, and the 1951 study insisted that a ballistic missile was feasible. The Air Force accepted Convair's judgment and MX-774 became Atlas. Subsequently, breakthroughs occurred in quick succession. Development was far advanced in 1957, and the missile reached the production stage in 1958.

Revival of the Air Force ICBM program was not premature. In 1952, Los Alamos was speaking not only of small atomic warheads but of small lightweight thermonuclear (TN) warheads as well. If the latter were married to a long-range rocket of reasonable accuracy, it could become an ideal strategic weapon. One year later the Teapot Committee, under Dr. John von Neumann, made a thorough study of ICBM potentialities and early in 1954 recommended that the program be accelerated immediately to take advantage of the new warheads. The Air Force reacted quickly. On 15 July the Air Research and Development Command (ARDC), as instructed by Headquarters USAF, set up the Western Development Division (WDD) at Inglewood, Calif., for the primary purpose of pushing the ICBM program.[24] Under Brig. Gen. Bernard A. Schriever, WDD--which subsequently became known as the Air Force Ballistic Missile Division (AFBMD)--was responsible for the management of Atlas research, development, and testing. In May 1955, WDD responsibility was increased to include the comparable, though more sophisticated, Titan.

It was good to have the Air Force ballistic missile program revived, and the Operation Castle thermonuclear shots in 1954 seemed to confirm the policy of developing rockets tailored to meet specific needs. Consequently, since TN warheads promised to be small, the Air Force sponsored correspondingly modest ICBM's. Though the missiles were capable of intercontinental range and possessed velocities great enough to achieve earth orbits if so desired, thrust was far less than that which the Soviets were developing for their missile-satellite programs.

Rocket propulsion received further impetus in 1955. In March 1954, at the President's request, the Science Advisory Committee (Office of Defense Mobilization) had established the Technological Capabilities Panel (TCP) to study the threat of surprise attack. Under the chairmanship of

Dr. James R. Killian, the panel, often referred to as the Killian Committee,* prepared an exhaustive report and submitted it to the President on 14 February 1955. Among other things, the report advocated an immediate program for intermediate-range ballistic missiles (IRBM's). The recommendation aroused some enthusiasm for various service proposals for IRBM's, and the interest heightened a few months later when NSC recommended that land- and ship-based IRBM's be considered essential to national security.[25]

In November 1955 the Secretary of Defense approved the Jupiter IRBM as a joint Army-Navy development task and authorized the Air Force to proceed with its own Thor IRBM. The Air Force promptly added this project to the other responsibilities of WDD. The Navy soon found that the Jupiter would be unsatisfactory for shipboard use and, late in 1956, obtained permission to withdraw from further participation in the project. In January 1957, the Navy gained approval of Polaris, a solid-fuel IRBM designed especially for submarine launching.[26] Though Polaris itself lacked intercontinental range, its mobility endowed the Navy for the first time with what amounted to strategic airpower. It gave the Navy a better position from which to argue roles and missions in the approaching space age, but Polaris missiles were not to be diverted from IRBM functions to serve as boosters in satellite launchings.

The following table lists the performance characteristics of the

---

*Other members were J.B. Fish, J.P. Baxter, J.H. Doolittle, L.A. DuBridge, L.J. Haworth, M.G. Holloway, E.H. Land, and R.C. Sprague, consultant.

missiles that proved to be important as "lift devices" for satellites and space vehicle projects between 1954 and 1959:

Missile	Alt or Range in NM's	No of Stages	Thrust Pounds	Velocity: mph	1st Successful Firing
Viking	154 (A)	1	21,000	4,000	1948
Aerobee	38 (A)	1	18,000	2,100	1949
Aerobee-Hi	164 (A)	1	18,000	4,600	1955
Redstone	175 (R)	1	78,000	3,500	1953
Sergeant	75 (R)	1	50,000	1,850	1956
Jupiter	1,500 (R)	1	150,000	10,000	May 1957
Thor	1,500 (R)	1	150,000	10,000	Sep 1957
Atlas	5,500 (R)	1½	360,000	18,000	Dec 1957
Titan	5,500 (R)	2 (1 of 300,000-lb thrust; 1 of 60,000)	360,000	18,000	Aug 1958

II. THE MILITARY SERVICES PLAN FOR SPACE, 1945-57

On the evening of 3 October 1942, that historic day when the first V-2 was successfully launched at Peenemuende, Walter Dornberger, by then a major general and director of the project, called together his chief assistants and said:[1]

> The following points may be deemed of decisive significance in the history of technology: we have invaded space with our rocket and for the first time--mark this well--we have used space as a bridge between two points on earth; we have proved rocket propulsion practicable for space travel. To land, sea, and air may now be added infinite empty space as an area of future intercontinental traffic, thereby acquiring political importance. This third day of October, 1942, is the first of a new era of transportation--that of space travel.
>
> So long as the war lasts, our most urgent task can only be the rapid perfection of the rocket as a weapon. The development of possibilities we cannot yet envisage will be a peacetime task. Then the first thing will be to find a safe means of landing after the journey through space.

The defeat of Germany was the end of Dornberger's work at Peenemuende, but the significance of rocketry for space travel was not lost upon those who had knowledge of the problem.

When the Russians went forward with their missile program in 1945, they included space travel as well as missile weaponry among their hopes. In the United States the Navy and Air Force--and eventually the Army--became interested in space but received little encouragement from the higher levels of government. Without support, the services could do scarcely more than sketch a program as something to be desired.

The Russian Space Program

The Russians handled their program with consummate skill. At the end of the war they made no secret of their intentions to conduct experimental missile work with the aid of the Peenemuende scientists imprisoned deep in the Soviet Union. In addition the Russians gave their own scientists adequate support to learn whatever the Germans had to teach and at the same time to go forward with their own experiments as rapidly as possible. Though they kept their program flexible, swiveled to meet the exigencies of international politics, the Russians did not deviate from the one unalterable aim of furthering the interests of the Soviet Union. Their method was scientific research in missilry, and their three chief objectives were to strengthen the nation's military prowess, enhance propaganda, and possibly in the end to prove the materialistic philosophy of communism by exhibiting life as universally indigenous to matter.[2]

As far as publicity was concerned, the Russian policy was "not to release any detail until we have experimental results" of a broad nature, said one Soviet official.[3] It was not until long afterwards that Western authorities learned that the Soviets undertook to develop a rocket-propelled intercontinental bomber in 1946; that they sent rockets to altitudes of 100 miles with payloads of 200 pounds in 1949; that their canine passengers of atmosphere research rockets were recovered by parachute in 1951; and that they conducted a systematic investigation of lunar-landing feasibility in 1953.[4]

Yet no Western statesman can plead a justifiable ignorance of the general nature of the Russian missile program. On 23 July 1945, <u>Life</u> published Peenemuende drawings of a large, manned space station, and there was every reason to suppose that the Russians had acquired similar or perhaps more advanced drawings of the same concept. It was common knowledge that rocket-propelled missiles and spacecraft were two aspects of the same research problem in the thinking of German scientists in the Russian prison camp. In October 1949, Karl T. Compton, then chairman of the Research and Development Board, wrote Louis Johnson, Secretary of Defense, and quoted an unidentified member of the board as saying:[5]

> Although reports from behind the Iron Curtain are meager, those which we have indicate that the Russians are exploiting the missile developments, which they inherited from the Germans, at high priority. They would be fools not to do so, now that the United States is so definitely committed to the atomic blitz. It would be tragic if we were to curtail our program now and let the Russians get ahead of us.

As time went on, it became the policy of Moscow frequently to remind the Russian people--and the world too, if it would listen--that the Soviet government was vigorously supporting a program to develop a space capability at an early date. The stream of information broadened appreciably in 1950--and was a flood by 1955. In 1951 there were several reports, emanating from government sources, that the Soviet Union envisioned a military space station and Earth satellite and that plans looked toward lunar landings within 10 or 15 years. This material could be found in many reputable journals in the United States.[6]

On 27 November 1953, A.N. Nesmeyanov addressed a "World Peace Conference" in Vienna. As the official representative of the Kremlin, he said that with available techniques it was possible to launch a satellite or send an object to the moon. Within days *Pravda* published the statement and discussed it with approving interest. In 1954 the U.S.S.R. Academy of Sciences established the Ziolkovsky gold medal for outstanding work in interplanetary communications. In April of the next year, the Presidium announced the creation of the Permanent Interdepartmental

Commission on Interplanetary Communications "to coordinate and direct all work concerned with solving the problem of mastering cosmic space." Simultaneously the Astronautics Section of the Central Aeroclub was organized "to facilitate the realization of cosmic flights for peaceful purposes." Even more indicative of a prospering space program were the appeals of Radio Moscow. Youthful volunteers were needed to help their country in its efforts to undertake flights to the Moon.[7]

Further evidence of Russian progress came in August 1955 during the Sixth International Astronautical Congress at Copenhagen, sponsored by the International Astronautical Federation. One of the two Russian "observers," L.T. Sedov, chairman of the Commission on Interplanetary Communications, declared that "it will be possible to launch an artificial Earth satellite within the next two years. The realization of the Soviet project can be expected in the comparatively near future." Sedov certainly did not speak without the knowledge and approval of his government, and his statement was tantamount to an official announcement.[8]

The Russians gave the world ample evidence of their space goal. Indeed the flow of information between 1951 and 1955 was a graph of Russian success, legible as a printed page at noon. In August 1958 the President of the United States approved an NSC statement to the effect that Soviet space accomplishments should long have been obvious to anyone, and a congressional committee concluded that "we did not need a spy system to give to the technically qualified a clear forewarning of Soviet progress."[9] It was surely a serious failure of the intelligence community if it did not warn the highest authorities of the looming crisis. It was an equally grave fault on the part of the highest authorities if they received but disregarded such a warning.

Service Moves toward Space Projects and Policy, 1945-48

The lack of an American space policy, or even a rocket policy, in the first few years after World War II compelled the separate services once more to shift for themselves in adjusting to the changing order. Aside from a keen interest in missiles as a form of artillery, the Army Ground Forces seems to have had no immediate awareness of space as a possible area of operations. The Navy and AAF/USAF felt quite differently about the feasibility of satellites, and they exhibited a realistic desire for space projects at the time when many civilian echelons expressed disapproval of such "impractical ideas."

In his final war report, 12 November 1945, General Arnold discussed the possible and probable use of new weapons in the future--projectiles, for instance, that might have a velocity of 3,000 miles per hour. In turn there would be new weapons of defense, and they would necessitate launching the projectiles nearer the target to give them a shorter time of flight and make their detection and destruction more difficult. Continuing, he said:[10]

> We must be ready to launch . . . /the projectiles/ from unexpected directions. This can be done with true space ships, capable of operating outside the earth's atmosphere. The design of such a ship is all but practicable today; research will unquestionably bring it into being within the foreseeable future.

Arnold's vision did not cause wide interest at the time, but before the end of the year the AAF negotiated a contract with Douglas Aircraft Company for a study of intercontinental warfare and its instrumentalities. On 1 March 1946, Douglas in turn organized Project Rand to fulfill these responsibilities. In this rather indirect way, therefore, it is possible to think of Rand, which became Rand Corporation on 1 November 1948, as having been established to investigate some of the possiblities of penetrating space.[11]

By the time of the AAF-Douglas contract the Navy, too, was interested in space, and it moved ahead of the AAF in formalizing a space program. Through the summer of 1945, Comdr. Harvey Hall, USN, and a few associates in the Electronics Division of the Bureau of Aeronautics (BuAer), worked with some of the material salvaged from Peenemuende. Hall suggested the desirability of a satellite test program, and on 3 October 1945 he was made chairman of a BuAer Committee for Evaluating the Feasibility of Space Rocketry (CEFSR).[12] Hall then opened discussions with Guggenheim Aeronautical Laboratory of the California Institute of Technology, Aerojet Engineering Corporation, Martin Company, North American Aviation, and Douglas. The talks led to contracts late in 1945 and early in 1946 with Guggenheim, Douglas-El Segundo, Martin, and North American for feasibility-study designs of space vehicles.

Within a matter of weeks CEFSR received several studies and chose the one from Douglas as the most suitable. The concept called for the development of a new rocket engine, using a fuel of liquid hydrogen and oxygen. The vehicle design employed the engine in clusters to obtain the desired thrust. The members of CEFSR were convinced that they had at hand a project of importance but one that would require general support in order to be approved by higher authorities.[13]

In March 1946, Hall approached the AAF representatives on the joint AAF-BuAer Aeronautical Board and broached the subject of a possible

Navy-AAF experimental space project. It was mentioned at a meeting of the board's Research and Development Committee on 9 April, and formal discussions were scheduled for the next meeting on 14 May. Prospects of a satisfactory understanding seemed good.[14]

At this point Headquarters AAF became interested in the negotiations from a policy viewpoint. The Air Staff agreed that if space were exploited the operations would be an extension of strategic airpower. Therefore the AAF should be the cognizant service. To avoid a possible compromise of the AAF position, it was essential to show a competence equal to that of the Navy. To Maj. Gen. Curtis E. LeMay, Deputy Chief of Air Staff for Research and Development, fell the task of safeguarding the AAF position. On 20 April 1946 he verbally requested Douglas-Santa Monica to have Project Rand personnel prepare a satellite feasibility study for the AAF. According to Douglas, the study was needed within three weeks "to meet a pressing responsibility." In light of this deadline the company assigned about 50 of its ablest scientists and engineers to the task, regardless of other activities.[15]

The study, "Preliminary Design of an Experimental World-Circling Spaceship," was written under great pressure. A first draft of 2 May showed inconsistencies that required overall revision, and on 12 May, Douglas officials hand-carried the paper to Headquarters AAF. There were shortcomings that could not be corrected within the time allowed, but none of these faults detracted from the superior quality of the study as a document of historical importance.[16]

Replete with pertinent formulas and tables, the study proved to be an engineering analysis of satellite feasibility. It showed conclusively that in 1946 American engineers using current techniques were qualified to begin work on a space vehicle that could have orbited a 500-pound satellite in 1951, six years before Sputnik. The first satellite would have orbited for 10 days or more before slowing to the velocity of re-entry, whereupon it would have been consumed by friction temperatures. Later versions, equipped with small wings and guidance control, could have been brought back to Earth and landed safely. One of the more important aspects of the study was its discussion of the advantages of liquid oxygen and alcohol fuel versus the possible use of liquid hydrogen and oxygen, a subject already treated in the Douglas-El Segundo study for the Navy.*

Admittedly the utility of a satellite could not be explicitly defined in 1946, but there were reasons to be optimistic. The time seemed not altogether different from the years immediately preceding the first airplane flight in 1903. The Wright brothers certainly had not foreseen fleets of B-29's bombing Japan or air transports circling the globe. But in 1946 it was possible to envision some of the scientific and military uses which a satellite could serve, and probably this appreciation of spaceship functions was far more accurate than the most realistic prophesies of airpower had been at the end of the nineteenth century. It was plain, said the report, that a satellite, being above the atmosphere, could make

---
* The controversy was an old one, and in general the problem of handling and storing liquid hydrogen had been regarded as offsetting the advantage of greater specific impulse. The Rand scientists followed the hypothesis that handling and storing would be solved in time. They therefore submitted design analyses for the structure and performance of rockets using both kinds of fuel. The hydrogen-oxygen-powered rockets showed an impressive theoretical reduction in gross weight, and the writers urged that "this /fuel/ combination should be given serious consideration in any future study." (Douglas Rpt SM-11827, 2 May 46, p IV.)

invaluable contributions to the study of cosmic rays, gravity, astronomy, the Earth's magnetic fields, weather forecasting, advanced methods of radio communication, space medicine, and interplanetary travel. Militarily a satellite could serve either as a reconnaissance craft, a guidance station to increase the accuracy of weapon-bearing missiles, or a missile itself to be brought down by remote control upon a chosen target. Finally, and perhaps most important of all for the immediate future of the country: "The achievement of a satellite craft by the United States would inflame the imagination of mankind, and would probably produce repercussions in the world comparable to the explosion of the atomic bomb."[17]

On 14 May 1946 the Research and Development Committee of the Aeronautical Board met as scheduled and began an interservice debate for which there could be no immediate conclusion. The Navy and AAF feasibility studies were seen to be eminently practical, each in its way, and the discussion turned not on which project should be accepted but on the much more difficult question of selecting a cognizant agent for military space activities. The AAF representatives insisted that service roles and missions gave the responsibility for intercontinental warfare to the AAF; that space operations would be an extension of that responsibility; and that by consequence the AAF was the agent of primary interest. The Navy argued for a joint Navy-AAF-civilian agency. Unable to agree, the committee sent the question to the Aeronautical Board where the consensus was that JCS would have to define roles and missions in space. The board then appointed a special subcommittee to its Research and Development Committee to coordinate space activities pending a JCS pronouncement.[18]

On 24 January 1947, after eight months of unrewarding discussion in the special subcommittee, the chief of BuAer appealed directly to the Joint Research and Development Board (JRDB) to create an ad hoc astronautics panel to coordinate, evaluate, study, justify, and allocate all phases of the earth satellite vehicle program. This move took the question to a much higher level and opened the way for the first statement of defense policy on space. The JRDB was directly responsible to the two service secretaries and could largely determine policies of research and development that were of joint interest to the Army and Navy.* Under the chairmanship of Dr. Vannevar Bush, the board operated through six committees--electronics, guided missiles, atomic energy, medical science, geographical exploration, and aeronautics. Although the committees, like the board itself, were under the chairmanship of distinguished civilian scientists, the membership was predominantly military except in the supporting administrative offices and technical panels, which were sometimes composed entirely of civilians.[19]

After JRDB received BuAer's appeal, the AAF protested that adequate coordination was already being done by the Aeronautical Board's special subcommittee. Three months later, to preserve its own authority, the Aeronautical Board requested that it be recognized officially as the agent to coordinate space projects.[20] There was no decision for several months. During that time, in September 1947, the National Security Act went into

---

*The Joint Research and Development Board was created by Robert P. Patterson and James Forrestal, Secretaries of the Army and Navy, who wanted a high-level agency to consider research and development policy. They decided to take the Joint Committee on New Weapons and Equipment away from the Joint Chiefs of Staff, reorganize it as JRDB, and make it responsible to the service secretaries. The first meeting of JRDB was held on 3 July 1946. (Minutes, 1st Mtg of JRDB, 3 Jul 46, w/statement of Hon Robert P. Patterson.)

effect; AAF was separated from the Army to become the United States Air Force; the Army, Navy, and Air Force were brought together within the National Military Establishment under the Secretary of Defense; and the Joint Research and Development Board became the Research and Development Board (RDB) with considerably more authority and power.

In December 1947 the Research and Development Board rejected the requests made by the Navy and the Aeronautical Board. At the suggestion of the Air Force, RDB directed its own Committee on Guided Missiles (CGM) to assume responsibility for coordination of satellite activities because "an Earth Satellite is considered to be a high altitude Guided Missile."[21] The transfer of satellite responsibilities to the CGM was not a happy omen for progress. Vannevar Bush, who continued to serve as chairman of RDB until 5 October 1948, discounted long-range ballistic missiles and satellites as "military dreams."[22] A conservative thinker and, judging by his own writings, of little imagination, Bush was nevertheless highly respected. He exercised wide influence in RDB and indeed throughout the executive branch of the Government.

Under the circumstances it is not surprising that the Committee on Guided Missiles showed little interest in the Navy and Air Force feasibility studies of May 1946. On 3 February 1948 the committee submitted the space question to one of its supporting agencies, the Technical Evaluation Group (TEG), with a request for recommendations before 31 March.*[23]

---

* TEG was composed of five civilians--Dr. W.A. MacNair, of Bell Telephone Laboratories, Inc., was chairman, and the other members were H.G. Stever, F.H. Clauser, R.W. Porter, and L.J. Henderson. TEG was established by CGM on 10 July 1947 to make technical analyses of progress in the field of guided missiles and assist CGM in formulating an integrated program. (CGM Directive, subj: Formation of a Technical Evaluation Group, 10 Jul 1947.)

The report, Satellite Vehicle Program, was ready by 29 March 1948. After considering the Navy and Air Force feasibility studies, the group concluded that both established the possiblity of a satellite, but the Navy's proposal would necessarily be the more difficult since it did not consider the use of existing techniques as the Air Force did. Moreover, since neither the Navy nor the Air Force established a military or scientific utility for a space vehicle "commensurate with the presently expected cost, . . . no satellite should be built until utility commensurate with the cost is clearly established." Also TEG noticed that the Air Force, still smarting from the reduced budgets of 1946 and 1947, was canceling contracts for ballistic missiles whose rocket engines might serve also as lift devices for satellites. To develop a special satellite rocket engine, as distinct from the engines of long-range missiles, would make a satellite program even more of a national luxury. The report therefore recommended[24]

> that the only activity directed toward satellite vehicles as such be a continuation of the Project RAND studies of the utility of such a vehicle. We believe, however, that the Navy and USAF should jointly sponsor and participate in these studies and that they should include such experimental work on auxiliary power plants, electronic apparatus, and the like as may be required.

Continuing, the report listed as desirable further studies of design and experimentation on multistage rockets; test-pit development of liquid-hydrogen motors; very lightweight tanks and structures to the point of full-scale static tests; guidance for long-range missiles; and hypersonic and superaerodynamic research.

In reply to CGM the three services concurred in the TEG report by midsummer with the understanding that the Army would continue research and development of short-range ballistic missiles; the Navy would proceed with the hydrogen-oxygen rocket engine being developed by BuAer as a possible satellite powerplant, but without fabricating a vehicle; and the Air Force would follow Rand studies on long-range rockets in order to determine the appropriate time to initiate development of complete vehicles. By these concurrences, the Army went far toward ruling itself out of space, the Navy continued to claim an interest in space, and the Air Force asserted its right to decide when a complete satellite vehicle should be developed. The CGM seemed to confirm the service positions on 15 September 1948 when it decided that Rand would continue its satellite studies, making them available to the Army and Navy as well as the Air Force, that each of the three services should continue its projects pertinent to space as mentioned in their concurrences, and that there be no other space activities for the time being. The committee then declared that its deliberations "on this item be considered closed."[25]

The Research and Development Board was then in a position to submit the following statement to the Secretary of Defense, James Forrestal, who included it in his First (1948) Report:[26]

> The Earth Satellite Vehicle Program, which was being carried out independently by each military service, was assigned to the Committee on Guided Missiles for coordination. To provide an integrated program with resultant elimination of duplication, the Committee recommended that current efforts in this field be limited to studies and component designs; well-defined areas of such research have been allocated to each of the three military departments.

For some time prior to and immediately after the RDB decision of September 1948 there were space activities that received no official recognition. For instance, the Martin company claims to have completed in 1947 detailed studies of a rocket vehicle that, had it been put on a crash program, could have placed a 1,000-pound satellite in orbit by 1949 or 1950.²⁷ Probably this report carried the Navy-Martin contract of early 1946 to completion. If so, it came to flower months too late. Martin complains that "nobody was interested." Nobody could be interested at the time because of current negotiations under way in the Aeronautical Board and the Joint Research and Development Board. After September 1948 the Martin report was obsolete.

Also in 1947 some employees of Army Ordnance at White Sands Proving Ground designed a space flight experiment that theoretically could have landed an object on the Moon. But the "stunt" was quickly shunted aside. On 24 February 1949, White Sands launched a Project Bumper missile* that soared to the surprising altitude of 250 miles. The event was only five months after the RDB decision, and Brig. Gen. Philip G. Blackmore, in command of White Sands, cautiously said in an interview that "the flight opens up new vistas for . . . exploration on the unknown regions of the atmosphere," but he did not mention space.²⁸

The cautious avoidance of public reference to space bespoke the hesitancy of the time to admit that the space age was already something to take into consideration. Probably the advent of the wheel, and certainly the coming of the screw propeller also excited disbelief and distrust, but caution had not prevented conquest of the land, sea, and air and could not prevent human ventures into space. The military services did not lose interest in space when RDB decreed caution in 1948. Indeed all three services had become involved in space projects of one kind or another long before the critical days of 1957. The Air Force especially was in a favorable position to continue probing the possibility of sponsoring a

---

*See above, p 15.

space program, and it took advantage of its opportunities.

The TEG report of March 1948 was not all that the Air Force could have desired, but as it was modified by service concurrences and approved by CGM and RDB, it was far from unacceptable. Indeed in September 1948 the Air Force could scarcely look back over the space negotiations of the previous two years without gratification. The timely Rand feasibility study in 1946 more than balanced the Navy's earlier effort and permitted the Air Force to argue with evidence in asserting its claim to be the service of primary interest in space. The rejection by RDB of the Navy's request for an ad hoc astronautics panel probably forestalled the establishment of a joint space agency to challenge the Air Force claim. The RDB acceptance of the definition of Earth satellites as high-altitude guided missiles bolstered the Air Force insistance that space weapons were strategic weapons. Moreover, RDB had authorized the Air Force to continue sponsoring Rand space studies, and these, combined with the pursuit of greater altitude, range, and speed, plus a growing knowledge of rocket potentialities, led inevitably to speculation of what space projects could be undertaken.

Resume of 1947-57 Rand Studies

Rand had not waited for RDB to give its decisions before continuing its work. On 1 February 1947 it released to the Air Force 12 new reports supplementing the original one, covering such subjects as building, launching, and orbiting of satellites.* In general, the reports clarified

---

*Reports prepared by Rand and released 1 February 1947:

1. RA-15021, Flight Mechanics of a Satellite Rocket.
2. RA-15022, Aerodynamics, Gas Dynamics, and Heat Transfer Problems

(contd)

thinking on space vehicles but did not define the utility of satellites. On 25 September 1947 the Chief of Staff, USAF, directed the Air Materiel Command (AMC) to evaluate the reports.

When AMC completed its evaluation of the studies in December 1947, progress in guided missile research by the Navy, Air Force, and others had reached the point where the actual design and construction of a satellite was already technically feasible, and some at AMC argued that a satellite project should be initiated at once. Prompt action would provide the necessary components by 1952. Lt. Gen. Howard A. Craig, DCS/Materiel at Headquarters USAF, was somewhat more conservative. He recognized the feasibility of satellites at the time but also that they were still economically too costly. This barrier, he felt sure, would gradually pass as technological progress brought down the cost of various components. He therefore urged the Chief of Staff to define the Air Force position as officially establishing an interest in space. This would guide lower echelons that might become participants.[29] On 15 January 1948, Gen. Hoyt S. Vandenberg, Vice Chief of Staff, signed the

---

(contd)
of a Satellite Rocket.
3. RA-15023, Analysis of Temperature, Pressure, and Density of the Atmosphere Extending to Extreme Altitudes.
4. RA-15024, Theoretical Characteristics of Several Liquid Propellant Systems.
5. RA-15025, Stability and Control of a Satellite Rocket.
6. RA-15026, Structural Weight Studies of a Satellite Rocket.
7. RA-15027, Satellite Rocket Powerplant.
8. RA-15028, Communication and Observation Problems of a Satellite.
9. RA-15029, Study of Launching Sites for a Satellite Projectile.
10. RA-15030, Cost Estimates for an Experimental Satellite Program.
11. RA-15031, Proposed Type Specification for an Experimental Satellite.
12. RA-15032, Reference Papers Relating to a Satellite Study.

following policy statement for the Air Force:

> The USAF, as the Service dealing primarily with air weapons--especially Strategic--has logical responsibility for the satellite.
>
> Research and Development will be pursued as rapidly as progress in the guided missiles art justifies and requirements dictate. To this end the problem will be continually studied with a view to keeping an optimum design abreast of the art, to determine the military worth of the vehicle--considering its utility and probable cost--to insure development in critical components, if indicated, and to recommend initiation of the development phases of the project at the proper time.

The Vandenberg statement followed close upon RDB's rejection of the Navy's request for a joint astronautics panel and its refusal to recognize the claims of the Aeronautical Board as the permanent agency to coordinate space activities. The statement antedated the TEG report by more than two months, however, and obviously served as the source for the Air Force's paper of concurrence reserving its right "to determine the appropriate time to initiate development of complete vehicles."[*]

During the next three years, Rand worked on satellite studies. The technique of orbiting a vehicle offered fewer theoretical difficulties than did determination of utility. Not until 1950 did Rand optimistically forecast the feasibility of a reconnaissance satellite. The next year a Rand report went so far as to advocate the development of such a vehicle carrying a payload of television equipment.[30]

On the strength of Rand's 1950 forecast, set forth in a "preliminary report," the Air Force gave its first space briefing to RDB, which proved

---

[*]There is also similarity between Vandenberg's comment that the military worth of a vehicle must be considered in relation to its "utility and probable cost" and the TEG report conclusion that neither the Navy nor the Air Force could show a military utility for a space vehicle "commensurate with the presently expected cost."

far more sympathetic than it had been in the earlier (1947-48) discussions. In sanctioning further USAF satellite studies, the action of the board was sufficiently positive to be interpreted as a confirmation of the Air Force as DOD-cognizant agent for space projects. The Air Force thereupon broadened Rand's space activities to include component research and design by industrial subcontractors.[31]

Another three years elapsed, 1951-54, during which the Air Force was concerned primarily with the Korean conflict, the expansion of the current military postures, and the revival of the ballistic missile program. At the same time, it preserved an expectant interest in satellites. In February 1954, Rand recommended that the Air Force develop a scientific satellite as a preliminary step to eventual utilitarian satellites, and in June 1955, in a supplemental report, Rand again urged the Air Force to support a scientific satellite because:[32]

> An artificial satellite circling the earth for days or weeks would provide information which cannot otherwise be obtained and which would enrich man's knowledge of the earth, the sun, and the universe to really unforseeable dimensions. Not to minimize the great contributions to science which the rocket program has already made, it may be permissible to say that it has allowed us only a glimpse of the unknown, showing the tremendous possibilities which would lie with a continuous observation station in outer space.*

No one in the Air Force would or could gainsay the desirability of a scientific satellite. The difficulty was that Rand's 1955 study recommended a project much more expensive than high-altitude research rockets without promising much more compensation than that already achieved at

---

*The 11 scientific uses of a satellite, according to Rand, follow: solar radiation measurements in ultraviolet and X-ray; electron density measurements; pressure, density, and composition measurements; cosmic ray measurements; albedo of the Earth; observation of meteors; measurements of the variation of the Earth's magnetic field; artificial seeding of the atmosphere; atmospheric drag measurements; goedetic measurements; and cosmic and solar hi-frequency radio noise.

White Sands. It was not easy for the Air Force to reconcile a costly scientific satellite with the need to preserve national security within budgetary limitations. Moreover, the Air Force was reaching out toward space at the same time with a useful reconnaissance satellite as one of its major objectives.

Air Force Space Projects, 1948-57

By the summer of 1957 the Air Force had four projects, with numerous subprojects, that aimed either to approach the fringe of or go into space with several types of aircraft, satellites, or other space vehicles. There was a research aircraft destined for new rocket-aerodynamic investigations and a more advanced boost-glide vehicle that, rising above the atmosphere, could serve as a spacecraft on reconnaissance or strategic bombing missions and then return to the atmosphere to complete its flight aerodynamically. In addition, there were plans for a reconnaissance satellite sufficiently versatile to fulfill several functions and a ballistic research and test system that could include lunar landings--if approved. Of course, behind all the plans for unmanned invasions of space was the belief that eventually manned spacecraft would carry human observers on space missions. To mention the conquest of space without assuming man's presence there would have been almost as unthinkable as the conquest of the sea without sailors, or the conquest of continents without adventurers to explore the new lands. Man-in-space was sometimes dismissed as a "stunt," but the concept remained as the conscious or unconscious raison d'etre of the space program, and the Air Force was already speaking of a man-in-space project by 1956. Simultaneously the Air Force continued to support the development of more advanced rocket engines.

Each of the space and near-space projects to which the Air Force was committed deserves a brief resume to indicate the extent of progress made between 1948 and 1957.

The X-15 Research Aircraft

As early as 1942-43 there were some in the AAF who foresaw that within a few years reaction propulsion in one form or another would bring airpower face to face with the hypothesis that no airplane could ever exceed the speed of sound because the velocity would pile up unpassable "air drifts" against the wings and under surfaces of the craft. It was as much to test the validity of the sound-barrier theory as anything else that the AAF initiated the X or research series of aircraft.[33]

In 1944 the AAF contracted Bell Aircraft Company to build the X-1. It was to be a glide rocket vehicle launched from B-29's at high altitude and then put under its own propulsion with rocket engines of 6,000 pounds' thrust, produced by Reaction Motors, Incorporated (RMI). The X-1 made its first powered flight on 9 December 1946 when it achieved a speed of Mach 0.9. Then on 14 October 1947 the X-1 broke the sound barrier, with Capt. Charles E. Yeager as pilot. Eventually the X-1 exceeded 1,000 mph, but by that time there was a more advanced aircraft available.

Learning the lessons of the X-1, Bell undertook the production of the X-2 and delivered it to the Air Force in 1952. After several accidents and delaying misfortunes, the X-2 exceeded 2,000 mph and reached an altitude of 126,000 feet. Meanwhile, the National Advisory Committee for Aeronautics (NACA), the Air Force, and the Navy contracted with Bell for the X-15, a rocket glide research aircraft intended to pass over into the near provinces of space. It would be equipped with an engine producing 50,000 pounds of thrust, and in 1957 there were expectations that the plane would eventually approach speeds of 4,000 mph and achieve altitudes of 50 miles or more. If these capabilities could be realized, the X-15 would enable the Air Force to test certain aspects of the principle of boost-glide.

Bomi--Robo--Brass Bell--Hywards--Dyna Soar

The Air Force was impressed by what it knew of Eugen Sänger's suggestion that the V-2 be used as second stage for a boost-glide vehicle that would be launched from Germany, rise above the atmosphere, and glide back as a very-long-range bomber against New York. In 1948, Rand spoke favorably of the idea, and some companies felt that it opened new areas of development. In 1952, Bell, where Dr. Dornberger was now employed, proposed a manned hypersonic boost-glide bomber/reconnaissance system that combined the Sänger concept with the more recent Rand studies.[34]

In 1954 the Air Force contracted Bell for a limited study of the boost-glide system. The conclusions were favorable, and on 12 May 1955 the Air Force issued General Operational Requirement (GOR) 92, which called for a hypersonic strategic bombardment system. The next year the Air Force called on Bell for a long-range boost-glide reconnaissance feasibility study. Bell called the bomber system Bomber Missile (Bomi), which was soon changed to Rocket Bomber (Robo). The reconnaissance study, which was kept separate from Robo, received the nickname of Brass Bell. At about the same time ARDC proposed the development of a boost-glide research vehicle called Hywards. On 30 April 1957, Headquarters USAF directed ARDC to consolidate Robo, Brass Bell, and Hywards into one project.[35]

During the spring and summer an ARDC ad hoc committee, which included representatives from NACA, Rand, and a few of the aircraft firms, worked on the assignment. The committee adopted a realistic approach of combining the three separate projects into a single program adjusted to a schedule that would permit cancellation by the Air Force with minimum cost should the concept prove impractical. The overall project, termed Dyna Soar, was organized as follows:

	Dyna Soar I (Hywards)	Dyna Soar II (Brass Bell)	Dyna Soar III (Robo)
1st Flight	1963	1966	1970
IOC[a]		1969	1974
Velocity	18,000 fps	18,000 fps	25,000 fps
Altitude	350,000 ft	170,000 ft	300,000 ft
Range		5,000 mi	Circumnavigation of the globe

[a] Initial Operational Capability

The ad hoc committee completed its work on 24 August, and ARDC presented the plan to the Air Staff on 17 October 1957. By that time the first Russian satellite had so changed the national outlook that Headquarters USAF directed ARDC to keep the project as described but telescope the schedule. Within weeks ARDC revamped the schedule:[36]

	Dyna Soar I	Dyna Soar II	Dyna Soar III
1st Flight	1962	1964	1965
IOC		1967	1968

The Advanced Reconnaissance System (ARS) or WS-117L

The boost-glide Brass Bell (Dyna Soar II), under the ARDC plan of August 1957, was assigned the theoretical range of 5,000 miles, and thus was not slated to be a satellite under its scheduled IOC date of 1969. However, out of the 1948-54 Rand studies had come recommendations for an

Advanced Reconnaissance System (ARS) project, which the Air Force approved.

Accepting as valid the 1950 forecast that a reconnaissance satellite was feasible,* General Putt, Director of Research and Development, authorized Rand on 19 December 1950 to enlist subcontractors to study and design several components--a nuclear auxiliary power unit, a television camera, an attitude-sensing device, and other items. Rand accepted this responsibility and called the project Man Hole (changed to Feed Back in 1952).[37] In May 1953, Headquarters USAF approved the subcontractors' work and directed ARDC to take over from Rand that phase of the work, which was then called the Satellite Component Study.†

In March 1954, Rand, still working on Feed Back, recommended an immediate high priority for a photographic reconnaissance satellite as one aspect of the project.[38] Headquarters USAF, with the approval of the OSD Coordinating Committee on Guided Missiles, then directed ARDC in August 1954 to proceed with development. Seven months later, 16 March 1955, Headquarters issued GOR 80 for a Strategic Reconnaissance Weapon System, and soon thereafter changed the name of Feed Back to Pied Piper.

For its part, ARDC went forward with the Feed Back/Pied Piper plans and tied the project to the Atlas missile for propulsion. At about the same time, ARDC announced that from October 1955 to April 1956 responsibility for ARS, or WS-117L as it was also being called, would be shared

---

*See above, p 41.

†Under the ARDC system of numerical designation, the Satellite Component Study was Project R-409-40. In 1954 the system of numerical designation was changed and Project R-409-40 became Project 1115.

jointly by Wright Air Development Center (WADC) and AFBMD. After April 1956, AFBMD would be responsible for managing the project.

Meanwhile ARS ran into two difficulties. First, the economy policy cutting research and development funds had crippled the project badly. The most valiant efforts of AFBMD, ARDC, and Headquarters USAF to win interest and support came to nothing. Worse, top officials within the offices of the Secretary of Defense and Secretary of the Air Force frowned on the project in the spring of 1955, even cutting back on those funds which otherwise would have been available. Thus, two years later, in the summer of 1957, development officials were still trying to excite interest in the project with tempting possibilities of combining ARS with Atlas and Titan missiles as lift devices to place payloads of 1,900 pounds in 300-mile orbits, or 1,100-pound payloads in 600-mile orbits. At that date, nothing could break the opposition of OSD. The Secretary of the Air Force showed academic interest but warned that insistence would create unfavorable repercussions at high political levels.[39]

Hypersonic Environment Test System

Considering that Dyna Soar and ARS could both trace back their lineage to the early years after the war, the Hypersonic Environment Test System (Hets) was a late comer among Air Force space interests. It had its origin in an ARDC proposal of 1956 that the United States should sponsor for scientific purposes a ballistic orbital and lunar research and test system.* The proposal advocated three phases: first, boosting a 200-pound payload to an altitude of 200 miles using Aerobee and Sergeant rockets;

---

*ARDC hoped to have high-level approval for the project as part of the contribution by the United States to the International Geophysical Year, scheduled for 1 July 1957-31 December 1958.

second, boosting a 500-pound payload to an altitude of 500 miles; and third, sending a payload into orbital flight to permit high-speed reentry studies.[40]

The Air Staff received the proposal with enthusiasm and concluded that the project could be expanded far beyond the original concept. On 31 July 1956, Headquarters USAF directed ARDC to revise and develop the plan into four phases that would include (1) boosting a test vehicle to an altitude of 300-500 miles; (2) boosting a test vehicle to an altitude between 1,000 and 2,000 miles; (3) combining the first two vehicles into a third that, using Atlas, would achieve Earth orbits and circumlunar flight; and finally (4) employing a vehicle of yet higher performance to permit lunar landings and interplanetary missions to the vicinity of Mars and Venus.[41]

ARDC rewrote the plan, called it Ballistic Weapons and Development Supporting System (Balwards) or WS-454L, and submitted it to Headquarters USAF on 15 March 1957. The Air Staff was gratified, but OSAF expressed opposition. Richard E. Horner, Assistant Secretary (R&D) informed the Air Staff that the project was too radical and must be rejected "at this time." Headquarters USAF therefore had no choice but to instruct ARDC to revise Balwards plans. Since satellite lunar and cislunar references were unpleasant, the third and fourth phases should be deleted.[42]

During the next seven months ARDC worked on the second revision, designated the Ballistic Research and Test System (Brats), and submitted it to Headquarters USAF on 18 December. The plan was for a long-range development, but by the end of 1957 the temper of the administration and the country required something quick regardless of significance. So Brats, like ARS, was long neglected. When it was revived late in 1958, it was redesignated Hypersonic Environment Test System (HETS).[43]

Man-in-Space (MIS)

In the early years of space thinking, 1945-57, no responsible person believed that spacecraft would take human passengers to the stars. Even with the most radical systems of propulsion, the journey's time would be too long. Yet no one conversant with the progress of rocket propulsion could doubt that within a few years men could be projected into space, either in satellites orbiting Earth or in spaceships traveling through some portions of the solar system. The question of man-in-space was, for the sophisticated, essentially a question of whether man could survive in the space environment once he got there. The success or failure of man to explore and possibly to conquer space depended upon a new science--space medicine.[44]

No science suddenly becomes part of human knowledge. It evolves gradually as one idea emerges from another like biological mutations in successive generations. Space medicine grew with easy transition from aviation medicine. As aircraft reached farther and farther into the heights above Earth, the human factor problems of high altitudes became the human factor problems of space. In this way, biologists, physicians, psychologists, and psychiatrists unwittingly began research in space medicine while handling the problems of flight in the upper atmosphere.

The purpose of space medicine was to learn how the space environment would affect the physiological and psychological behavior of human beings. In 1945 no one could predict the cardiovascular and respiratory effects of weightlessness; the exact danger of ambient radiation; the reaction of the human body and its perceptual and decision-making functions to the high G's of launching and the vacillating G's of reentry. Nor could anyone say whether the tangible loneliness of life in a space vehicle would be psychologically bearable. Answers could be found only through patient research.[45]

Research in very-high-altitude environment began in 1946 when the Aeromedical Laboratory at Wright-Patterson AFB and the National Institute of Health decided to participate in the White Sands upper-atmosphere experiments. They called on Holloman AFB, N. Mex., located near White Sands, for local support. Eventually, for the sake of convenience, Wright-Patterson established the Aeromedical Field Laboratory at Holloman. By late 1951 overall planning for the work was a responsibility of ARDC's Director of Human Factors.

The first task of the Air Force-Institute of Health group was to find and master the techniques of sending live specimens into space and effecting their safe recovery. At an early date some of the instrumented nose cones that replaced the V-2 warheads bore fungus spores and fruit flies to detect the effects of cosmic radiation. By 1948 small animals were sent aloft in Aerobee capsules specially designed to control temperature and pressure. Many of the experiments were annulled by takeoff accidents, and even more were lost through faulty recovery methods. It was not until 1951 that a monkey was successfully launched and returned. Nevertheless, much was learned between 1946 and 1951 from electronically gathered evidence about the behavior of animals at high altitudes.[46]

Despite the great accomplishments of space medicine between 1946 and

and 1951,* the most successful projects were frequently ridiculed by heavy-headed commentators. Their criticisms were neither a credit nor a help to the United States, but their words of disapproval matched the national policy of economy. Space medicine lagged from 1951 until late 1957. Little was learned during those years except from the research aircraft of the Air Force and the Navy† and from some unobtrusive high-altitude balloon flights.[47]

The X-1 and X-2 operations showed clearly by the early 1950's that it was time to think of means to sustain life in advanced models and boost-glide vehicles, plans which were under way. Progress was made in designing pressure suits, but little was done to provide a habitable cabin, a prerequisite for journeys into space of more than a few hours.[48] It was in connection with this necessity that the balloon flights were most helpful.

Between 1952 and 1955, Holloman's Space Biology Laboratory accomplished 78 successful ascents. Numerous small animals--hamsters, mice, and dogs--went to altitudes of 100,000 feet or more, remained there for several hours, and thereby tested the lethality of cosmic radiation. Results convinced the scientists that the danger was less than anticipated, and plans began for Project High Man, the use of balloons to take human passengers to equal altitudes and remain there for a day or more. The

---

*In 1951 the School of Aviation Medicine, Randolph AFB, Tex., held the first major international meeting on the subject of space medicine. It was a symposium on "The Physics and Medicine of the Upper Atmosphere." Distinguished scientists came from all over the world, and the published proceedings remained the standard reference work in the field for a number of years.

†The X-1, X-2, and D-588-I.

project required larger and more elaborate capsules than those ever used before. The first High Man flight occurred on 2 June 1957. Capt. Joseph W. Kittinger reached an altitude of 95,000 feet and floated there for hours, his capsule, instrumented for 25 experiments, serving as a space laboratory.* The second flight, on 19 August 1957, carried Maj. David Simons to 102,000 feet, and the balloon remained aloft for more than 32 hours. These experiments proved the "adjustability" of man to the space environment if provided with a habitable capsule.⁴⁹

By 1956 the progress made in space medicine, the evidence acquired from the balloon experiments, and the promise of AFBMD to have Atlas and Titan ICBM's operational within a few years began to fit together nicely. In February of that year, ARDC proposed that the ICBM's be modified to accommodate a man-inhabited capsule for orbiting, just as the V-2 had been modified with nose cones for small life. Recovery was still the most serious difficulty. In March the Air Force approved plans for a Manned Ballistic Rocket Research System and stirred interest in several aircraft companies. In December both Avco Corporation and Martin submitted unsolicited proposals. Others soon followed, and by April 1957 the Air Force would have contracted for a ballistic capsule study had adequate funds been available. Shortly thereafter the situation changed rather radically, and that which had been "last" became "first" in national interest. Believing that the time was auspicious, ARDC proposed on 1 November 1957 that a group directed by the Aeromedical Laboratory undertake

---

*The Navy had previously established its Stratolab, but for long periods this project remained inactive. Also, the High Man experiments were not conducted by Air Force agencies alone; the Navy supplied the helium, and the Army sent two helicopters for tracking. In a limited way, High Man became an interservice project.

the development of a "life support capsule" as a "subsystem" nose cone in an ICBM. The method seemed the quickest, simplest, and least costly way of getting man in space.[50]

Propulsion

As space projects moved forward from speculation to feasibility studies, and from feasibility studies into research and development status, the Air Force became increasingly interested in more advanced types of rocket engines than those current within the missile program. In 1955, prophets of space spoke, for the most part, of Viking, Aerobee, Aerobee-Hi, Redstone, Sergeant, Jupiter, Thor, Atlas, and Titan.* By the summer of 1957, space propulsion requirements were obviously coming to exceed those of the missile program. At that time WADC could thankfully mention, in addition to the 12 current liquid-rocket engines of interest or possible interest to Air Force space projects, 4 other liquid-engine development and study projects already under way as well as 8 solid-engine projects.†

---

*See above, pp 13-24.

†The following lists were presented in a SAB ad hoc committee briefing on 29 July 1957 by Ezra Kotcher, Directorate of Laboratories, WADC:

12 Available Liquid-Rocket Engines		4 Liquid Rockets in Dev & Study	
System	Thrust Pounds	300,000-lb LOX-JP rocket engine	
		150,000-lb IRFNA-UDMH rocket engine	
Navaho		75,000-lb Nuclear-rocket study	
LR83-NA-1	415,000	1,000,000-lb rocket-engine study	
LR71-NA-1	240,000		
Titan	300,000	8 Solid Rocket-Engine Projects	
Titan sustainer	55,000		
Atlas	300,000	System	Thrust Pounds
Atlas sustainer	60,000		
Thor	150,000	X-17	50,000
Redstone	75,000	Q-5 booster	48,200
X-15	50,000	Snark booster	132,500
Bomarc	35,000	F-100D launch	130,000
Hustler	15,000	State of art	250,000
Rascal	12,000	Goose	38,400
		Matador booster	101,000
		2d-stage IRBM	100,000

The variety and therefore the choice of engines, either available or under development and that could be used for a space program, was surprisingly great.

The list, when considered in the light of Air Force contributions to the development of Aerobee-Hi and Navaho, is indicative of the debt that the Army and Navy, and indeed the nation as well, owe to USAF pioneer efforts in the field.*51

The First Army-Navy Space Project

Almost simultaneously with the Air Force decision in the summer of 1954 to proceed with Project Feed Back, which so soon became Pied Piper and by 1955 was WS-117L or ARS, the Army and Navy proposed jointly the development of a satellite. It was the first time the Army had come forward to claim a foothold in space, and it was the first attempt on the part of the Navy since the RDB decision of September 1948. Undoubtedly the Air Force did far more between 1948 and 1954 to promote a space program than the Army and Navy, and by 1954 there were very respectable Air Force space projects being considered. Nevertheless, by one of the ironies of history, the Army-Navy proposal in 1954 was of more immediate promise than anything the Air Force could offer because it depended upon the use of off-the-shelf components, produced as part of the Army's ballistic missile program.

---

*Kotcher's list of "present" and "future" rocket engines also included both the Project Rover nuclear-rocket study and the 1,000,000-pound engine study. Both of them belonged far into the future, or so it seemed in 1957. Project Rover was a USAF-AEC attempt to determine by 1961-62 the feasibility of a nuclear-rocket powerplant. Study on the 1,000,000-pound-thrust engine had only recently begun, and though its eventual importance could not be questioned, especially in the dawning space age, it was not immediately significant.

In 1954 there was a growing interest in "scientific satellites," and Wernher von Braun propitiously suggested that the Army undertake the project. It would have been relatively simple and inexpensive to put together a vehicle from on-the-shelf hardware of the Ordnance Department and then launch it with a Redstone missile. After some consideration, Army headquarters decided that it would be advisable to make the project a three-service undertaking. The Navy accepted, but the Air Force was already too deeply interested in getting the reconnaissance satellite under way. The Army and Navy together therefore worked out during 1954 and the early part of 1955 a scheme to place a 5-pound inert slug in orbit as a scientific project to prove the feasibility of satellites. The Redstone missile was selected as the booster with three upper stages of clustered Loki rockets. The project became known as Orbiter, and the launching date was set for 1956.[52]

Thus, as early as 1954, the Army, Navy, and Air Force were all actively engaged in sponsoring space projects. But attempts by the armed forces to explore space were disapproved by the national administration whose space-for-peace policy aimed to keep space free of military intrusion.

III. A POLICY OF SPACE-FOR-PEACE--AND ITS EFFECTS

Foreign policy and technology have always been closely related. Sometimes the relationship has been positive as when governments have challenged technology to produce new and secret weapons in the interest of national prestige and security. Sometimes the relationship has been negative in the sense that statesmen have sought to interdict new weapons or restrict the cost of armaments in the interest of peace or economy.

To stay within the twentieth-century history of the United States, technology of the early 1900's gave President Theodore Roosevelt the Great White Fleet as a "big stick" to calm the troubled areas of the Pacific and elsewhere. By 1914, technology had done much to bring Europe to the verge of war, and in May of that year President Woodrow Wilson tried to avert the coming crisis by asking for an end to the arms race. In 1921 the cost of technologically modern navies was so great that President Warren G. Harding hitched American policy to a naval moratorium as a substitute for collective security. In 1940, President Franklin D. Roosevelt called on American technology to supply the nation with 50,000 aircraft annually to curb Nazi power through intimidation or in battle. During his two administrations, 1945 to 1953, President Harry S. Truman relied upon the technology of nuclear weapons to thwart Soviet imperialism, and he was largely successful in doing so.

By 1955 the space age was incontrovertibly at hand, and President Eisenhower undertook a unique maneuver. For the first time in history he attempted to exclude militant imperialism from a locale that was still technically inaccessible to man. He enunciated a space-for-peace policy that would have excluded warcraft from the areas in which aerodynamic vehicles could not operate. This policy obviously had a profound effect on the course of the American space program. Whether in the long view of history the policy would be named wise or unwise, it constituted the intellectual medium in which the program took shape during its early years.

In brief, the space-for-peace policy was the frame of reference for many of the program's critical decisions between 1955 and 1959.

The First Concepts of Space Law

Spaceflight is inherently international. The phrase was used in testimony before the House Select Committee on Astronautics and Space Exploration in the spring of 1958, but the idea was far from novel. Long before Sputnik, legal scholars expressed the same thought in different words, but until spaceflight became theoretically possible, space was nationally and internationally meaningless. It was with the first perception of rocketry as a practical means of space propulsion that space became a possible field of international rivalry and conflict.[1]

The international implications of spaceflight were so obvious, indeed, that the V-2 rockets in 1944 raised conjectures of possible complications that might arise from the future use of long-range, high-altitude missiles. The surmisings seemed unrealistic for the most part, however, until 1951 when John Cobb Cooper, member of the Institute for Advanced Study, Princeton University, gave substance to theory in an erudite paper on "High Altitude Flight and National Sovereignty."[2] Thereafter many creditable articles appeared throughout the world, and classified documents on the same subject began to accumulate in government files.

Despite minor differences, there was general agreement on fundamentals among scholars in the West, but not on how the fundamentals could be obtained. If space was to be saved from the chaos of national rivalries, it was necessary to determine the extension of sovereignty in altitude, define airspace and outer space, and establish tests for the legality of future ownership of celestial bodies. With unwonted optimism the legal specialists turned to history for guidance, but the appeal was not helpful. Few worthwhile signposts were at hand.

The two important treaties on flight through airspace--the 1919 Paris Convention Relating to the Regulation of Air Navigation and the 1944 Chicago Convention of International Aviation, neither of which was signed by Russia,[*] recognized "the exclusive sovereignty" of each state in the airspace above its territory. Neither convention contained provisions directly applicable to outer space, since the term was not then within the vocabulary of international law. It appeared only that sovereignty, until arbitarily limited in the future, would extend to indefinite distances beyond the earth, projected upward either by parallel or radial verticals of national boundaries.[3]

Either technique would impose serious obstacles on the exploration of space. Under the circumstances the most pressing requisite was to define airspace and outer space because the line between those two areas seemed most likely to be acceptable as the highest altitude of sovereignty.

---

[*]The Russians were not asked to subscribe to the Paris Convention. In 1944 they were invited to the Chicago Convention. They accepted the invitation, and their delegates left Moscow by air, but while they were flying over Canada, approaching Chicago, they were recalled home without explanation. (House Hearings before the Select Cmte on Astronautics and Space Exploration, 1958, p 1281.)

The proponents of space law then turned back to the history of maritime law for precedent and analogy. They pointed out that the law of the sea evolved with the rise of nationalism. The age of discovery and the Spanish-Portuguese-English rivalries of empire led first to unilateral claims to the sea, or to the large areas of the sea, as part of national domains. Later the claims lapsed when the increasing number of strong nations prohibited enforcement. From this situation came the doctrine of the freedom of the seas, and this in turn was slightly modified by the seaward extension of sovereignty within the range of coastal defense, or as justified by other considerations. The principle was crystalized by the latter part of the eighteenth century in the "three-mile limit," and it generally persisted thereafter despite occasional attempts to extend the distance. It was not unreasonable to hope that international agreements would recognize some specific distance above the earth as analogous to the three-mile limit, beyond which there would be freedom of space comparable to freedom of the seas.[4]

Along with these discussions many writers recognized that sooner or later there would be the question of legality to space claims as there had been to territorial claims in the sixteenth and seventeenth centuries. From the time of Columbus and his immediate successors it was customary for explorers to witness the colonial claims of their monarchs by leaving upon the shores of new lands figures of the crown and cross as emblems of sovereignty and state religion. The claims were further strengthened by taking back to the homeland small quantities of soil and a few branches of vegetation. A yet stronger claim came with the establishment of colonial settlements, and in the twentieth century the Permanent Court of

International Justice decreed in the Norwegian-Danish dispute over Greenland that there must be also "the exercise . . . of sovereign authority." Similar methods of imposing ownership on celestial bodies would probably follow landings on the moon and planets,* and, indicative of the spirit of the twentieth century, several writers assumed that the first earth visitors to astral realms would leave scientific instruments as symbols of national claims.†5

Such was the thinking among the experts in international law on the subject of spaceflight between 1950 and 1955. The fact that reputable scholars were becoming concerned with the problem was noted in the Department of State, and this doubtless convinced some authorities that the time had come to give official thought to the international significance of space.

First Efforts to Create a Space Law, 1955-57

Looking back with the wisdom of hindsight, it is easy to see that the years 1954 and 1955 were critical in the history of the world. It was then that not only were decisions made on man's first ventures into space but Soviet and American space policies were determined that directly affected the formulation of space law.

The compelling force behind the earliest space project sponsored by the American government was the plan of scientists to hold the International

---

*There were suggestions that a happier alternative would be for international law first to provide that, like Antartica, celestial bodies would be subject to no one sovereign authority.

†When Lunik II reached the moon on 13-14 September 1959 it planted on the lunar surface metal pennants inscribed with the name and coat of arms of the Soviet Union. However, the Russians made no colonial claims to the moon at that time.

Geophysical Year (IGY). There was a long but thin history behind the scientific program. The scheme was an outgrowth, or perhaps an expansion, of previous International Polar Years of which the first had occurred on 1 August 1882 to 1 September 1883, when 48 nations in the vicinity of the Arctic Circle studied simultaneously and reported on various phenomena. The results were interesting and encouraging, and a second International Polar Year was undertaken during 1932-33, commemorating the fiftieth anniversary of the first. A third international endeavor was slated for the end of the next half century, 1982-83, but such great scientific strides were taken between 1933 and 1953 that scientists were unwilling to wait for another 30 years. In the midst of preparatory conferences, the concept was broadened to become an International Geophysical Year, and the time agreed upon was 1 July 1957 to 31 December 1958.[6]

In October 1954 a special committee of the International Council of Scientific Unions (CSAGI) met in Rome. Here an almost crucial decision was made to include among the IGY activities the launching of small satellites for scientific purposes.[7] The determination of IGY scientists to explore space may have come as a shock to some officials in Washington, because it meant that the issue of the international significance of space could no longer be ignored. The United States and the Soviet Union were clearly the only nations capable both financially and technologically to support the original experiment. Undoubtedly the Russians would gladly offer their cooperation to IGY authorities, and the United States could scarcely afford to do otherwise.

On 15 April 1955 the Soviets announced the establishment of its Special Commission for Interplanetary Communications. The meaning of the move was plain. Since the commission was given the responsibility of designing and producing "a remote control laboratory to circle the Earth as a satellite,"[8] the Russian statement was tantamount to an announcement of a satellite program, and a boast that the work was already far advanced. Concurrently, outstanding Russian scientists spoke confidently of the Soviet program. Military rockets would be harnessed to place a satellite in orbit. Later, other satellites would circle the moon, and these in turn would be followed by radio-sounding spaceships and eventually by manned vehicles.[9] The Soviet program was admirably bold, and its remarkable success in the next few years was impressive evidence of straight thinking in Moscow.

Simultaneously an American program, taking shape in high-level deliberations in Washington, was being delineated with circumspection. Since a satellite circling the earth must unavoidably pass over foreign territories, it was necessary, thought the President and the Secretary of State, to impress upon the world that American space vehicles were peaceful. In February 1955 the President received assurance from his top scientific advisers that no satellite as then conceived could be employed as an offensive weapon. If the vehicle released a bomb it would not fall upon the territory below, but would continue circling the earth in the wake of the satellite.[10] Here was a clear distinction between aircraft of high altitude and satellites. It followed that, as a matter of defense, the sovereignty of a nation should extend upward through the area navigable by aeronautical craft, but above that height the area should be accepted as free of national boundaries because it was not amenable to offensive weapon systems. If the point could be universally accepted in 1955, it might serve as the basic canon in an international law prohibiting combat above the atmosphere regardless of future technological progress in space weaponry.

The next question confronting the President was the kind of propulsion to be used by the satellite. The Russians had stated that they would employ military missiles to orbit their satellites, and the President wanted to put that decision in contrast with American aims by rejecting the use of military missiles to penetrate space. He sought the advice of his highest political advisers, and in May 1955 they agreed that the American satellite should be orbited by nonmilitary rocket engines.*[11] Neither President Eisenhower nor his advisers feared the delay their idealism would impose upon the American satellite project by requiring the development of a special "civilian" booster.

The President could have announced the satellite project at once but apparently refrained for diplomatic reasons. In May and June he was preparing for a "summit meeting" in Geneva, scheduled for 18-19 July. It was there he presented his "open skies" proposal to the Russians. His move began a persistent effort on the part of the United States to tie in space exploration with disarmament and the creation of an international law that would keep space altogether free of military rivalries.[12]

The Russians showed little enthusiasm for the President's proposal. The exchange of military blueprints was not likely to appeal to a nation as imperialistic as the Soviet Union. Equally annoying, no doubt, was the implication that they modify their frequently asserted claim to absolute sovereignty in the airspace above their homeland. On this point they

---
*See below, p 71.

remained adamant until success with their space program made it advantegeous for them to argue differently.[13]

The President was doubtless disappointed by the Russian rejection of his inspection proposal. It meant also that he could not use the American satellite in immediate negotiations with Moscow for the recognition of the freedom of space. The time had come, regardless of Moscow, to announce publicly that the United States would launch a series of small satellites entirely peaceful in nature, as one of the contributions the nation would make to IGY. On 29 July 1955, one week after the President's return from the Geneva conference, an official statement came from the White House that the United States was indeed undertaking a satellite project for scientific purposes. And, to prove the good intentions of the United States, the satellite would be launched by a specially developed nonmilitary rocket engine.[14]

So it was the public learned of the first government-sponsored scientific satellite. On 1 August 1955, Nikita Khrushchev took advantage of a reception at the Swiss legation in Moscow to express his willingness, and the willingness of his nation, to "support" the American space effort if the interests of humanity could thereby be served.[15] These were fine words, but Russo-American cooperation in space projects was more easily envisioned than achieved.

In the autumn and late winter of 1955-56 the United States conducted a meteorological study that entailed the lofting of balloons from many locations, including some in West Germany and Turkey. When the balloons passed over Soviet territory the Russians protested vigorously. They claimed that the hidden purpose of the balloons, as well as the open-skies proposal, was to obtain photographs needed to make maps of Russia. The United States denied intentions of espionage, but on 7 February 1956 the Department of State decreed that no more balloons should be released. Secretary of State John Foster Dulles said that he acted as a matter "of decent friendly relations." He added, almost as a warning, that the most reasonable interpretation of international law made the ownership of airspace and outer space "a disputable question."[16]

The balloon incident showed that high-altitude flights were fraught with international complications, and there was no reason to suppose that the Russians would be more kindly disposed toward satellite overflights of their territory, no matter how peaceful the satellites might be. Moreover, the President's advisers were no longer as sure as they had been in February 1955 that space and space vehicles were without military significance. The President was cognizant of these changing ideas and became even more devoted to a space-for-peace formula. In his State of the Union message on 10 January 1957, he expressed a willingness to accept an international agreement to control reliably "the development of missiles and satellites." Again, he linked together his space-for-peace with his hopes for disarmament.[17]

Four days later, Henry Cabot Lodge, the American ambassador to the United Nations (UN), presented a more detailed version of the same plan to the General Assembly. The Russians made no direct reply. But in March and April, Soviet representatives argued for the prohibition of nuclear weapons, and they interpreted the term to include rockets of any range if equipped with nuclear warheads. In this way the Soviets made space control dependent upon the elimination of all nuclear weapons. Such a prohibition, in Western opinion, was already unenforceable however because the Russians had rejected Eisenhower's doctrine of the open skies and all other forms of effective inspection.[18]

On 25 July 1957, Harold Stassen, the President's Special Assistant for Disarmament and the American representative on the Disarmament Subcommittee of the UN General Assembly, reiterated the need to establish control over experimentation with objects traveling through outer-space--this meant apparently both ballistic missiles and satellites. He warned that the situation was perilously close to that of 1945-46 when, following the Hiroshima-Nagasaki bombings, the rejection of the Baruch control plan had led to an international race for atomic weapons. Stassen hoped the same mistake would not be made in the development of space vehicles, which involved an equal, and perhaps an even greater, danger for mankind. He proposed a technical committee of the world's eminent scientists to devise an inspection and detection system that would guarantee the peaceful uses of space.[19]

Space politics in the United Nations did not go beyond that point prior to Sputnik, but it had gone far enough to show the positions of the United States and the Soviet Union.

IV. LOST OPPORTUNITIES--VANGUARD AND OTHERS

Long before 1954 it was common knowledge that aeronautics, the science of creating and operating aircraft, was limited by a ceiling. Air-breathing engines would not operate above an altitude of approximately 140,000 feet. It was equally well known that rocket engines had no such ceiling. Rockets developed a momentum that depended entirely on Newton's third law of motion and were independent of the atmosphere as an oxidizer for combustion, thereby removing all limits to the altitudes attainable. Finally, there was no law of physics to restrict either the rocket's thrust or payload capacity. The rocket could be used to deliver warheads against earthbound enemies or to propel vehicles into the depths of the solar system and beyond. If Peenemuende and White Sands had any meaning it was that astronautics--the science of designing, manufacturing, and launching of spacecraft--was inevitable. If the Navy and Air Force feasibility studies of 1946, and all the subsequent plans, had any meaning, it was that the techniques of astronautics were rapidly being mastered. These facts were abundantly clear when the Army and Navy proposed Project Orbiter in 1954.

Nevertheless there were some who, even if they saw the inevitability of space travel, could not see its importance. On 19 November 1954, Von Braun warned that the possessor of the first space station would be in a position to rule Earth. The next day Secretary of Defense Wilson was asked in a news conference if he agreed with Von Braun. "No," said Wilson, "I would rather keep my feet on the ground, figuratively speaking as well as physically speaking. I don't know that anyone knows how you would rule the world with a space station. It is a little dreamy, I think." Two weeks later Wilson was reminded that the Russians might orbit a satellite before the Americans. "I wouldn't care if they did," he replied.[1]

The President's Decision to Support a Space Project

Even as Secretary Wilson argued that space was a paradise for dreaming scientists, the situation was changing. It was in March 1954 that the President, to guard against a second Pearl Harbor, created the Technological Capabilities Panel under the chairmanship of Dr. James R. Killian.* In its report, submitted to the President on 14 February 1955, the panel dealt with many subjects, among them the importance of space vehicles in the near future as instruments of intelligence.[2]

The Killian Committee had a thorough understanding of the technical difficulties and possibilities of exploring space. The report pointed out that large surveillance satellites would have to await the development of ICBM rocket engines, for nothing else at the time could supply the required booster thrust. On the other hand, small satellites weighing 5 to 25 pounds could be orbited by Redstone engines, which would soon be available in quantity. The total cost would be moderate. In what was doubtless a reference to Orbiter, the report stated optimistically that a "project of this kind has been proposed by the Department of Defense, and may already be underway." In its concluding remarks on the subject, the panel declared: "The new prestige that the world will accord the nation first to launch an artificial earth satellite would better go to

---
*See above, pp 22-23.

the U.S. than to the USSR."

On 11 March 1955 the Assistant Secretary of Defense (R&D) transmitted the Killian Report to the service secretaries. He requested a JCS position on the report that he could forward to the National Security Council, already engaged in determining the kind of satellite program the United States should support. The services were gratified that the panel indorsed the concept of a reconnaissance satellite. It seemed to sanction the Air Force Feed Back and the Army-Navy Orbiter plans and indicated that these projects had "a general alignment favored by the country's highest scientific talent".[3]

It was while the Joint Strategic Plans Committee (JSPC) studied the panel's report that the Navy, on 23 March 1955, officially requested OSD approval of Project Orbiter. The administration was just then in the midst of formulating its space policy, which was already being premised on the space-for-peace thesis. The Secretary of Defense, acting in light of the current trend, promptly quashed the Navy's proposal. On 28 March he expressed his disapproval of Orbiter, and his directive to the three service secretaries was so phrased as seemingly to include Feed Back as well:[4]

> Because of important policy questions involved, these departmental programs must be carefully considered and fully coordinated. The Assistant Secretary of Defense (Research and Development) is assigned responsibility for such coordination. Further funds will not be committed for work in this area without his prior approval.

Wilson's memorandum was discouraging, but it did not deter the JCS on 18 April 1955, in its comments on the panel report, from asserting a military need for a surveillance satellite. The Joint Chiefs added that to be useful the satellite would have to be much larger than the one being considered by the Government.[5] Presumably the Secretary of Defense sent the JCS comments, or at least an abstract of them, to the National Security Council, still engaged in satellite deliberations.

The JCS opinions, however, were of little influence. On 26 May 1955, NSC expressed its confidence in a space-for-peace policy. Though acknowledging the necessity of a space project, the council ignored the JCS requirement for a surveillance vehicle. It called for the development of a satellite divorced from military significance and lifted into orbit by a nonmilitary booster. Thus, NSC determined the nature of America's first space venture and cast aside the Orbiter project.[6]

The President approved the policy statement on 27 May. Following a delay of two months, while futile negotiations were under way with the Russians over open skies, the White House announced on 29 July 1955 that the United States would launch a series of small, purely scientific satellites in the course of IGY. The military, for the sake of efficiency, would have only managerial authority in contracting with industry for the design and production of the satellite components.

Specifications for Vanguard

In a sense the White House announcement was premature. Although the May policy statement settled the type of satellite to be developed, the means of propulsion and the managerial agency within the Department of Defense remained unsettled. Discussions, under the direction of Assistant Secretary of Defense Quarles, were proceeding among the three military services and various important committees, but decisions were still pending.

In view of the responsibility assigned to him by Wilson's directive of 28 March, Quarles had turned at once to the Coordinating Committee on General Sciences for advice and guidance, stating that the project would be a triservice effort but tied to IGY commitments. He also specified that the satellite itself would be unclassified, although the means of delivery could be classified.[7]

The committee submitted its report on 4 May 1955. It expressed confidence in the feasibility of the satellite, urged continuation of the study, and suggested that each of the three services prepare satellite proposals within the broad outline already determined. At once Quarles directed each of the three services to submit plans and soon thereafter created the eight-man, all-civilian ad hoc Advisory Group on Special Capabilities--sometimes called the Stewart Committee for its chairman, Dr. Homer J. Stewart*--to consider and evaluate them.[8]

Three weeks before the White House announcement of 29 July the three services submitted their separate plans. The Army brought back Orbiter with only minor modification. The Navy too urged Orbiter but, fearful of its rejection because it called for the use of a military missile as booster, suggested a backup plan based upon the use of Viking, the test vehicle produced several years before for upper-atmosphere research. The Viking was free of military connotations, and its thrust promised to lift a sphere 20 inches in diameter and weighing 20 pounds into an orbit with a perigee of 150 to 200 miles.

---

*Other members of the group were: Drs. C.C. Furnas, R.R. McMath, C.C. Lauritsen, John B. Rosser, Richard W. Porter, Joseph Kaplan and G.H. Clement; Athelstand F. Spilhause as alternate; and Paul S. Smith and Joseph C. Meyers as secretaries.

The Air Force faced a dilemma. Although vitally interested in the exploration of space, it could do only one of two things, neither of which seemed likely of acceptance: propose the use of Atlas or the development of a new nonmilitary rocket engine, either of which would almost certainly interfere with the general progress of the ICBM program. In the end the Air Force submitted plans for Project World Series and urged employment of Atlas as the booster. The Air Force was thus practically ruled out of responsibility for the nation's first space program by circumstances and the Administration's prejudice against the use of a military missile.[9]

The characteristics of the three service proposals can be briefly summarized in tabular form:[10]

	Army	Navy	Air Force
Booster	Redstone	Viking	Atlas
2d stage	Loki	Aerobee-Hi	Aerobee-Hi
3d stage	Loki	Aerobee-Hi	
4th stage	Loki		
Thrust at sea level	78,000 lbs	27,000 lbs	330,000 lbs
Cost	$18 million	$20 million	$16 million
Ready date	Late 1957	Mid 1957	Early 1958

On 4 August 1955 the ad hoc advisory group sent its report to Quarles. There was a reasonable assurance that the United States could put a small scientific satellite in orbit during IGY. Admittedly, Atlas would give the greatest performance margin and permit the largest payload. However, the group also thought that the Air Force plans would interfere with the ICBM program, and this involved "points of national policy outside the competence of the group." As between the Army and Navy plans, the group voted five to two in favor of Viking. Here again the advisability of

employing military boosters influenced the decision. The use of Redstone would create problems of security and, since Redstone facilities and manpower were limited, might prove disadvantageous both to the missile and satellite programs. Also, from a technical viewpoint, Viking required only two additional stages whereas Redstone required three. The minority favored Orbitor because Redstone was larger than Viking, had fewer development problems, was already entering flight-testing, and therefore would have the benefit of many tests before the time of satellite-launching.[11]

With this report in hand, Quarles sought the advice of the Research and Development Policy Council, composed of the three service assistant secretaries (R&D) and high-ranking development officers of the Army, Navy, and Air Force. Quarles was chairman. The council concurred in the recommendations of the advisory group, thought not unanimously. Army representatives insisted that Orbiter was the better plan since it depended upon the proved components of Redstone and was more likely to succeed than Viking. On 15 August the Army warned the Secretary of Defense that, because of time-consuming development requirements, the Viking plan might enable the Soviets to launch the first satellite, an event of incalculable effect on American prestige.[12]

OSD chose to ignore the Army's warning and on 9 September 1955 approve the Viking plan, soon to be known as Vanguard. OSD also instructed the Army and Air Force to cooperate with the Navy, under whose management the project would be developed. Actually, the Navy served as project manager with authority to contract with industry for the necessary components. Simultaneously, OSD warned the three services on 19 September that they could not develop any other satellite of their own.[13]

Approval of the Viking plan meant that the prime contract went to Martin, who had first designed and produced the research vehicle. Another contract went to General Electric to modify the Aerobee-Hi's 20,000-pound-thrust Hermes engine into a 27,000-pound-thrust first stage for Vanguard. Aerojet was to adapt the Aerobee main engine into a Vanguard second stage. Either the Grand Central Rocket or the Alleghany Ballistics Laboratory would design the third stage.[14]

The Unnecessary Delay

Rocket authorities considered the Vanguard concept to be technically excellent. Had it been approved two or three years sooner, it would have sufficed to meet the temporary national needs. Undertaken as it was late in 1955, the competitive element of the United States forging ahead of the Soviet Union's space program made Vanguard a risky venture. To be successful, "something had to be done within 2 years that had never been done before in 2 years."[15] Even so, Vanguard might have come through on schedule had it not bogged down in prejudice. The Department of Defense did not consider Vanguard a project of "first importance" and allowed only a "dribbling release" of requisite funds.[16]

There was protest, within the Department of Defense and elsewhere, but it did not overcome the dominating indifference. In mid-August 1955, Quarles replaced Harold Talbot as Secretary of the Air Force, and he brought to his new office the same caution that characterized his work as Assistant Secretary of Defense (R&D). On 22 November, Clifford C. Furnas, chancellor of the University of Buffalo, succeeded Quarles as Assistant Secretary of Defense. He was a scientist of repute and had served on the Stewart Committee. He was therefore highly qualified by training and experience to appreciate the requirements of Vanguard. There was hope that he might succeed in breaking through the wall of indifference. He failed, and resigned in protest. He later blamed Wilson for the "financial congestion" that held back Vanguard in spite of warning that Russia would succeed in putting the first satellite in orbit. Furnas said that Wilson adopted a "so what" attitude toward the program and sidetracked Vanguard funds when they were most needed.[17]

Lower military echelons, and interested civilians too, became alarmed by the program's slowdown. True, the Vanguard first stage was ready for firing on 8 December 1956, and in the next five months there were six other firings, all of them successful. In every case, however, the second and third stages were dummies. This was a serious matter because the success of Vanguard was dependent on all three stages. The situation became more grave because of the mounting evidence that the Russians were preparing to launch their satellite at an early date. In June 1957, F.J. Krieger of Rand predicted the first Soviet satellite would be launched in the late summer or early autumn, suggesting 17 September as a probable date because it would mark the centenary of Ziolkovsky's birth.[18]

During these same critical days there was much high-level haggling over the cost of Vanguard. The original Navy estimate had been $15 to $20 million. The total rose steadily, and in January 1957 the Bureau of the Budget estimated that it would be $83.6 million. Arrangements had been made to fund $70 million, which left $13.6 million still to be provided. In April 1957 the Bureau of the Budget reestimated the cost, raising it to $110 million, which left $40 million to be funded.[19]

There were sharp arguments within the Government on the advisability of continuing the project. The Bureau of the Budget and the National

Science Foundation were reluctant to invest more money. The Vanguard proponents argued that the program could obtain information of importance for missiles, especially on micrometeoric matter; that the scientific community of the world would be shocked by such a retreat; and that cancellation would vitally affect the prestige of the United States. The President turned to the National Security Council for advice. At its meeting of 10 May 1957 the Council gave Vanguard a reprieve. The project could continue but without further elaboration. Indeed, if possible, the cost should be cut.[20]

Last Possibilities for a U.S. "First"

The summer of 1957 was a period of anxiety for those who understood the situation and dreaded the consequences of a Soviet "first" in space. Until 4 October they hoped that either Washington would approve an Army project somewhat akin to Orbiter or that the Air Force's controversial Project Far Side would succeed. Neither hope was realized.

Among the experts of the Army Ballistic Missile Agency (ABMA), some felt that the slowdown in Vanguard might persuade OSD to reconsider the rejected Orbiter plan, in part at least. The 1954 proposal to develop a satellite project along with Redstone had the touch of reality.* After the selection of the Viking in 1955, ABMA continued its regular experiments, and these included the further development of the Jupiter-C. This was a multistage missile based on the Redstone but intended as a test vehicle for the Jupiter program.[21]

---

*Work on the preliminary designs for Redstone started in 1950. Progress was rapid after approval in the spring of 1951, but the first Redstone operational unit as a field weapon system was not ready until May 1958.

The Army attempted several times to obtain permission to use the Jupiter-C as a means of orbiting a satellite but in May 1956 was expressly forbidden by OSD to do so.[22] At least the Secretary of Defense did not forbid continuation of the Jupiter-C, and it was fired for the first time on 20 September 1956 as a three-stage vehicle with the Redstone as the first stage. It lifted an 84-pound payload to an altitude of 680 miles over a range of 3,300 miles. Van Braun and his associates realized at once that had the 84-pound payload been replaced by a fourth stage* it could have gone into orbit. The date of 20 September 1956 therefore marked the existence of an American capability to place a satellite in orbit, but the Government did not take advantage of its own resources.[23]

The uneasiness of the Army increased in the summer of 1957 when the Russians announced their development of long-range missiles, threatened to use them in the Suez crisis, and in August demonstrated their possession of an intercontinental ballistic missile. It was evident that the Soviets had reached the frontier of outer space and were preparing to launch a satellite. The tragedy of the situation, as seen by the Army and its missile team at ABMA, was later summarized by Lt. Gen. James M. Gavin, then Deputy Chief of the Office of Research and Development: "We had the scientists and the industrial facilities to keep ahead of the Russians. The failure was in decision-making."[24]

Though outwardly the Administration showed no change of heart, there were rumors of uneasiness at high levels. Reports of this concern came through to military field agencies and to some interested civilians. One periodical expressed its belief as late as July that something would be

---

*This is what was later done for the first Explorer satellite.

done to permit the ABMA team to show its competence:[25]

> Eulogized and advertised ad nauseum as mankind's greatest adventure there is still no assurance that any of the VANGUARD launchings attempted will be successful during the 18 months of the IGY. It's the nature of the still-young state of the rocket art.
>
> Even if VANGUARD is ready by spring it still may not be first. Reports point to a Russian try within 10 weeks. And to the south the /Army/ missile team /at Huntsville/ everybody tries to ignore may beat even that date.

Thus, until almost the last minute there was expectation that the Administration would call on the Jupiter-C to outmaneuver the Russians. But days passed, and the word that could have given America primacy in space did not pass from Washington to Huntsville.

There was yet one more opportunity for the United States to achieve a spectacular though not a satellite success before the Soviets could capture the imagination of the world with Sputnik. Project Far Side was the Air Force version of a concept that slowly evolved from proposals made in 1951 by Dr. S. Fred Singer, of the University of Maryland, for further research in the upper atmosphere.[26] By 1954, Singer was talking about a minimum orbital unmanned satellite of earth (Mouse) project, which expanded the upper-atmosphere research from instrumented high-altitude rockets to satellites. The idea won the interest of Col. William O. Davis, Air Force Office of Scientific Research (AFOSR), ARDC, whose enthusiasm equaled his great determination.[27]

During the next two years there were widening discussions of Mouse. Then, in 1956, Davis and Morton Alperin, also of AFOSR, attended the annual international astronautical conference in Rome. They heard the Russian representative speak of the Soviet plans for a satellite, and, knowing that Vanguard moved slowly, came to the conclusion that Mouse, if properly supported, might serve as an American balance to the coming Russian success. With further thought, Davis drew up a modified version of Mouse. It called for the firing of six rockets in an upper-atmosphere research project intended to gather at an altitude of 4,000 miles information of "vital interest to the Air Force."* To overcome the drag of the lower atmosphere the rockets would be fired from balloons at an altitude of 100,000 feet.[28]

For a number of reasons, Far Side became controversial. If properly supported, it would require perhaps as much as 7 percent of ARDC's limited research funds for fiscal year 1957. There was also the question of jurisdiction. Since Far Side, if approved, would be a geophysical experiment, it rightfully belonged to the Air Force Cambridge Research Center (AFCRC) rather than AFOSR, but Davis questioned the depth of AFCRC interest in the project. He therefore faced a dilemma. He could either abandon the project or proceed so quietly--some would say "furtively"-- that opposition would fade in the ignorance of what was being done. He chose the second approach, and until 15 March 1957, Lt. Gen. Thomas S. Power, commander of ARDC, was not briefed on the subject.[29]

By that time so much had gone into the project that, as Davis had foreseen, it was difficult to withdraw. Despite charges of subterfuge,†

---

*The information gathered would pertain to magnetic fields, cosmic rays, and the propagation of radio signals in extreme altitudes.

†For instance, coordination was not always open. The project was only listed in AFOSR in July 56 as "Status of Research Proposals," but in September was listed as "accepted." When challenged on this point, Davis said coordination was normal for an "unsolicited exploration research proposal." (Ltrs, B/Gen H.F. Gregory, COMAFOSR to M/Gen W.M. Morgan, COMAFCRC, 5 Apr 57, & Morgan to Gregory, 17 Apr 57; ltr, Col W.O. Davis to COMARDC, 25 Oct 57, subj: Coordination of Project Far Side.

mismanagement, and "utter misdirection of basic research funds," ARDC requested the Air Staff to appeal to the Secretary of the Air Force and through him to the Department of State, the Department of the Interior, and the Atomic Energy Commission to permit tests at Eniwetok.* Permission was granted in June 1957.[30]

The first shot was fired at the end of September. It was a failure. The balloon, carrying the rocket to be launched by radio, went up a few thousand feet and then suddenly fell into the sea. The second attempt, on 4 October, was a near success. The balloon rose to 90,000 feet and then began a slow descent. When it was down to 70,000 feet the crew made a last-minute attempt to save the day and fired the rocket. Though it became entangled with the collapsing balloon, the rocket was traced to an altitude of 370 miles. After that the instrument was silent.†[31]

* * * * *

The next morning, newspapers around the world bannerlined the 184-pound Russian Sputnik, the first manmade satellite in history.

---

*Actually Far Side survived only because of the contractor, Aeronutronic, Inc., which duplicated the money allocated by the Air Force. In June 1957 six balloons were delivered by Aeronutronic from which the rockets could be fired, and the crew moved overseas for testing. At the same time Davis was relieved of his duties at AFOSR and was succeeded by Col. Eugene LaVier. Despite the high qualifications of Colonel LaVier, this shift reportedly injured the morale of Far Side personnel.

†In the third test, on 6 October, a short in the firing mechanism triggered the rocket, which could not be traced. In the fourth attempt, the balloon froze at 56,000 feet and shattered. The fifth attempt was on 19 October. The rocket was fired at 96,500 feet but was damaged in passing through the balloon and sent back few signals. The sixth and last shot was on 22 October. Its signals were heard for eight minutes, which meant that the rocket penetrated between 2,500 and 4,000 miles into space. Brig. Gen. H.F. Gregory pointed out in justification of Far Side that, despite all the criticisms that could be made, had the operation succeeded, it would have offered a spectacular success to which the Administration, the Department of Defense, the Air Force, and the American people would have been glad to point on the unhappy morning of 5 October 1957.

V. A POST-SPUTNIK RESHAPING OF POLICY

Sputnik marked a magnificant and historic advance in science. As such, it deserved the congratulations that the President of the United States gave the Soviet government on 9 October 1957. No American would have felt other than kind envy had the first satellite been orbited by a friendly power. But coming as it did from Communist Russia, dedicated to the "burial" of free man, the triumph created dismay everywhere outside the iron curtain. As a congressional committee phrased it: "We face the terrifying prospect that nuclear attack upon the United States can be directed from Soviet bases." In addition, there was the new challenge to America's preeminence in the world of technology, the loss of international prestige, and the fact that Russia had staked out for herself primacy in space.[1]

In contrast with an early tendency toward "hysteria"--for so the first American reaction has been described--the shock and surprise of Sputnik had some beneficent effects. The American "smug sense of superiority was shattered," and out of the national humiliation came a more calm realization that, among other things, there had to be a reexamination of foreign and domestic policies on questions of space projects, defense organization, strategy, and the desirability of a civilian-scientific space program that would far exceed the little ambitions of Vanguard. Again quoting a congressional committee, there was widespread admission that America's misfortune was attributable only to indifference in the past on the part both of the people and the Government:[2]

> Soviet Russia's ability to develop atomic and hydrogen weapons so soon after the United States did, should have been warning enough to galvanize our national efforts. Our intelligence of Soviet missile experiments should have hoisted higher the red flag of danger. But until the American people read about, and could see for themselves if they cared to look, a luminous metal ball revolving in the heavens, Russian progress in science and production was serious discounted.
>
> Now the American people must respond to the fact that we have a great and powerful rival in the most complicated technical and industrial fields. They must respond, not in panic, not in diffuse and wasteful motion, but in a calm and purposeful dedication to the task of building up the nation's strength. Our country must be strong and unexcelled in the weapons of war; it must use that strength in the difficult, but unremitting, search for peace.

Once the chagrin of Sputnik had somewhat subsided, the press demanded and the Government attempted a judicious appraisal of the situation. In the area of foreign affairs, the President decided upon, and Congress approved, a reassertion of the pre-Sputnik space-for-peace policy, but qualified to accommodate a very restricted military program and a very ambitious civilian-scientific program. So the United States came to sponsor a three-fold space policy--international, military, and scientific. The three-fold policy itself underwent considerable change between October 1957 and the end of 1959, but always space-for-peace came first, and to that end the military program remained subordinate to the civilian.

At First, Emotional Reactions

Throughout the American press and seemingly throughout the foreign press as well, the first reaction to Sputnik was expressed in sharp criticism of the Administration. Editorials in the United States especially condemned "the partial measures, hit or miss planning and confused organization that have marked our . . . work in this field."[3]

Opinion on Capitol Hill was caustic in general, and the unfavorable comments were not limited to representatives of one party. Senator Symington warned that the position of the Free World would soon become intolerable unless strong remedies were introduced by the Administration without delay. Senator Henry M. Jackson regarded Sputnik as a "devastating blow to the prestige of the United States." Senator Styles Bridges said it was time for Americans "to be less concerned with . . . the height of the tail fin on the new car and to be more prepared to shed blood, sweat, and tears if this country and the free world are to survive."[4]

At the level of the White House and the Cabinet there was a tendency, said Newsweek, for officials "to hide behind the pretense of being undisturbed." Presidential Assistant Sherman Adams spoke of the accomplishment as "outerspace basketball"; James Hagerty, presidential press secretary, said Sputnik was unimportant because it had not caught the President unawares; soon-to-be-retired Secretary of Defense Wilson said the Russians had performed a "neat scientific trick." On 9 October 1957 a White House press release announced cryptically that the United States would not engage in a space race--and that the Vanguard schedule would not be accelerated. The statement was fat with unconcern. Yet again, on 3 November, the 1,120-pound Sputnik II, complete with dog, was casually dismissed by Hagerty as being "no surprise to the President."[5]

The press generally interpreted the Sputnik belittlement policy as a sign of nervousness, and the interpretation was not altogether without supporting evidence. Between 8 and 15 October the President and his advisers held numerous conferences "looking toward a re-evaluation of the missile program"--a comment that perpetuated the confusion in many minds of missiles with satellites. In late October there was a mistaken report that the third shot of Far Side had penetrated 4,000 miles into space. Without waiting for verification, the Department of Defense embarrassingly hailed the "achievement" as proof of a vigorous program in research and development. The press could also note that after Sputnik II the President called further conferences on the subject of the missile program.

Despite the President's assertion that the Vanguard schedule would not be accelerated, there were signs of acceleration. The Vanguard schedule had called for several test vehicle shots before attempting to fire a genuine satellite vehicle. The first test vehicle shot was scheduled for early December 1957, and by November the Administration fastened upon this event and inflated it to portend the actual launching of a satellite—an undertaking for which NRL lacked the opportunity for adequate technical preparation. Of course, on 6 December the shot failed to orbit and the United States was again humiliated unnecessarily.

By the end of the year the nation was beginning to accept the unpleasant fact that the space program had lagged too long to catch up with the Russians in the near future. The President was more reassuring than he had ever been before when, on 9 January 1958, in the course of his State of the Union message, he said quite simply that "most of us did not anticipate the intensity of the psychological impact upon the world of the launching of the first satellite."

Meanwhile, between the appearance of Sputnik I on 4 October 1957 and the President's message to Congress on 9 January 1958, much thought had been given to the space policy that the United States would pursue in the future.

A Modified Space-for-Peace Policy

The flights of Sputnik around the earth brought into sharp focus the earlier academic question of sovereignty in space and its violation by space vehicles. There was still no line of demarcation of areas to be closed or open to international traffic, whether military or civilian-scientific.

With perfect aplomb, the Russians protested after 4 October 1957 that no one could accuse them of violating the rights of other nations by the satellite moving overhead. Sputnik had not passed over any foreign territory; it was simply that foreign nations passed under the orbit of Sputnik. Along with this casuistry, the Russians unofficially proposed in periodicals that an international agreement should limit sovereignty to an altitude of 12 miles, or at most 18.[6] The Americans were less precise in their comments. For instance, Von Braun said there were no exact division lines in nature. The question of sovereignty would have to be settled by arbitrary decisions, and he suggested, for no particular reason, that an altitude of 100 miles could be accepted as the division between national sovereignty in altitude and the free space from which military vehicles might be prohibited. He added that 300 miles or 1,000 miles would be equally acceptable.[7] Rear Adm. Hyman G. Rickover, Assistant Chief of Bureau of Ships for Nuclear Propulsion, USN, was more military in his approach. He said:[8]

> The dividing line between military and civilian uses /of space/ could arbitrarily be set at the . . . maximum permissible altitude for a missile of 12,500 miles . . . . The distance of 12,500 miles is the maximum distance a missile would be required to travel. The earth being 25,000 miles in circumference, 12,500 miles is the maximum distance between any two points on earth.*

---

*The statement was reasonable in 1957-58, but by 1960 it appeared that anti-missile defenses might make it necessary for the attacking missile to be sent toward its target the long way round, a distance that could far exceed 12,500 miles.

He added that delineation would reserve for each nation an area "analogous to the 3-mile limit for international waters." Such a boundary for national rights would not only insure ample altitude for IRBM's and ICBM's but would be high enough to permit the employment of some military satellites as well.

Both Von Braun and Rickover deserved and received the highest respect, but neither of them spoke for the U.S. Government. The President, of course, was the arbiter of policy, and in reaching his decisions, he was guided by his immediate advisers, especially by Secretary of State Dulles and by leaders of Congress.

During all the months of pre-Sputnik effort to define and support a space-for-peace policy, the Government had not committed itself on details of space law or limits of sovereignty. The President, between 1955 and 1957, did nothing more officially than seek an international agreement, through the United Nations, to limit the exploration of space to peaceful purposes and to tie this in if possible with a move toward disarmament. Consequently, in October 1957 the President was free to continue negotiations for a space law without being hamstrung by previous commitments. However, since the United States did not immediately protest the flight of Sputnik above American territory, this silence could be interpreted as a tacit admission that all space was free for scientific exploration since the Russians claimed that function for Sputnik.[9]

The position of the President and the Secretary of State during the last few weeks of 1957 and far into 1958 seems to have been that for the immediate future the United States should do no more than continue the effort to negotiate agreements to keep space for peaceful purposes and to countenance a space program at home demonstrating the nation's peaceful

intentions. At the same time, since circumstances demanded some form of
military program, the President could justify it as necessary pending the
realization of international control. In doing so, he could appeal to
the charter of the United Nations, to which the Russians had subscribed,
for legal justification. Article 51 of the charter recognized the right
of a nation to defend itself against attack from any direction, a provision
as applicable to space as it was to land, sea, and air.[10]

The President received support in maintaining such a position from
Congress even at the time when both houses were attempting to insure an
adequate space program for the Department of Defense. Throughout the
first six or seven months of 1958, many senators and representatives expressed individually their approval of the Chief Executive's space-for-peace policy. In June the Senate and House passed a concurrent resolution
"that the United States should strive through the United Nations" to reach
an international agreement "to banish the use of outerspace for military
purposes, provide for joint explorations of outerspace, and establish
methods to settle disputes which may arise." The resolution had the support of both the Department of State and the Department of Defense and
seemed to express a complete agreement between the executive and legislative branches of the Government.[11]

After the exchange of American and Russian views in the General Assembly in January and April 1957 and Harold Stassen's statement before the
Disarmament Subcommittee of the General Assembly in July,[*] the United
Nations took no further action on space for several months. Then, shaken
from their lethargy by Sputnik on 4 October, 20 nations joined with the

---

[*]See above, pp 66-67.

United States on 11 October in bringing before the General Assembly a draft disarmament resolution that called for the peaceful uses of space. Without attempting definitions, the resolution implied that "outerspace" meant the region above and beyond the farthest altitude at which the atmosphere could hinder the orbiting of satellites. Also, the silence of the sponsoring nations on the subject of their sovereignty, like the silence of the United States, could be interpreted as a concession that space beyond the atmosphere was free.

In another one of the long intervals of patient waiting engendered by the vast machinery of the United Nations, the Russian and American positions were made clear in a direct correspondence between President Eisenhower and Nicolai A. Bulganin, nominal Prime Minister of the Soviet Union. On 11 December 1957, Bulganin proposed a summit meeting on disarmament. On 13 January 1958 the President replied, urging again that disarmament begin by limiting the use of space to peaceful purposes. Eisenhower warned that both the United States and the Soviet Union were "using outerspace" for testing missiles designed for military purposes. He thereby admitted that IRBM's and ICBM's followed trajectories that made them space weapons. This renewed tie-in of space vehicles with missiles opened the way for Bulganin to reply on 3 February that an agreement to use space only for peaceful purposes could be reached without difficulty if the Western powers would ban fission and fusion weapons altogether and liquidate foreign bases.[12] So the argument was back where it had been 12 months before.

On 25 March 1958 the new Soviet ambassador to the United Nations, Valerian A. Zorin, supported by the solid bloc of Communist state representatives, requested the General Assembly to include on its agenda an item to prohibit the use of "cosmic space" for military purposes and to call for the withdrawal of all troops from bases held in foreign countries. The United States countered on 2 September by requesting international cooperation in space to parallel progress in disarmament. This sparring was simply a repetition of old arguments. The Soviets wanted to use space as a means of eliminating American military bases in Europe; the Americans wanted to hold the bases pending an effective international control of space exploration.[13]

On 17 September the General Assembly compromised by placing the Russian and American proposals under the single heading of "Questions of the Peaceful Use of Space" and submitting them to its First Committee (Political and Security) for consideration. Debate began 12 November 1958 and moved back and forth along the well-trodden arguments.[14]

Meanwhile the administration in Washington had determined upon new tactics. On 18 September, Secretary Dulles addressed the General Assembly and urged the prompt creation of an ad hoc committee to speed agreement on the creation of a permanent agency. Ambassador Lodge repeated the request in the First Committee on 13 November, and 19 other nations supported the proposal.* At the same time, possibly to appease the Russian delegates, Ambassador Lodge rephrased the American policy on space and disarmament by urging that the study of space should proceed regardless of any other questions. He hoped that agreement on the peaceful use of space might reduce international tensions and the need for armament.[15]

---

*The 19 nations were Australia, Belgium, Bolivia, Canada, Denmark, France, Guatamala, Iran, Ireland, Italy, Japan, Nepal, Netherlands, New Zealand, Sweden, Turkey, the Union of South Africa, the United Kingdom, and Venezuela.

The Russians showed no sign of cooperation, and the President decided to demonstrate the solidarity of American opinion in backing the Administration. He turned to the joint congressional resolution of the previous June and requested Lyndon B. Johnson, majority leader of the Senate, to support the Administration's space-for-peace policy by addressing the General Assembly and affirming the nation's unity on the subject.[16] Speaking in New York on 17 November, Johnson said that the Congress of the United States had requested the President to appeal to the United Nations for international cooperation in space. He assured the General Assembly that there were no differences "within our Government, between our parties, or among our people" on the need to keep space for peaceful exploration. He urged that there be no differences among the 81 UN members. He concluded:[17]

> Today, outerspace is free. It is unscarred by conflict. No nation holds concession there. It must remain this way. We of the United States do not acknowledge that there are landlords of outerspace who can presume to bargain with the nations of the earth on the price of access to this new domain. We must not--and need not--corrupt this great opportunity by bringing to it the very antagonisms which we may, by courage, overcome and leave behind forever through a joint adventure into this new realm.

The address was effective, and the Russians indicated a willingness to cooperate with the Americans in preparing a joint resolution without reference to the military bases. Direct conversations between Lodge and Zorin raised hopes of settlement. On 24 November, however, Lodge announced that though there was agreement on the need for an ad hoc committee, there was disagreement on membership. The Soviets stood for an 11-member committee to include Czechoslovakia, Poland, Rumania, and the U.S.S.R. The Americans argued for representation in accordance with interest in space; the Russians argued for a proportional representation by bloc. The First Committee sanctioned an 18-member plan, and the General Assembly approved the decision on 13 December. The Czechs, Poles, and Russians, joined by the representatives of India and the United Arab Republic, refused to participate because the Rumanians were excluded, and the 18-member committee thus became a 13-member committee in actuality.* It began work in the spring of 1959 and submitted a report on 26 June that solemnized the usual platitudes and urged the creation of an autonomous organization to deal with space problems.[18†]

The 14th General Assembly convened in September 1959 and began consideration of the report. Christian Herter, Secretary of State since 22 April, addressed the Assembly much as Dulles had done a year before and asked the Russians to cooperate; Kuznetsov, again the ranking Soviet representative, responded favorably and proposed creation of the Committee on the Peaceful Uses of Outerspace as a permanent agency. On 12 December the Assembly established this committee with representatives from 24 states.[19‡]

---

*Argentina, Australia, Belgium, Brazil, Canada, France, Iran, Italy, Japan, Mexico, Sweden, the United Kingdom, and the United States.

†The committee also emphasized the coordination of radio frequencies for tracking, communications, and research purposes as the "first technical area in which immediate international action was required, suggesting the International Telecommunication Union (ITU), a UN agency, as a means of handling the problem. The ITU, with representation from 80 nations, met at Geneva in August-December 1959, but little was done. The United States also focused attention on the World Meteorological Organization (WMO) and urged that it study the use of meteorological satellites. As a consequence, WMO established a special panel in 1959, with the United States as a member. (House Hearings before the Cmte on Science & Astronautics, 86th Cong, 2d Sess, Review of the Space Program, pp 28-32.)

‡Albania, Argentina, Australia, Austria, Belgium, Brazil, Bulgaria, Canada, Czechoslovakia, France, Hungary, India, Iran, Italy, Japan, Lebanon, Mexico, Poland, Rumania, Sweden, the Soviet Union, the United Arab Republic, the United Kingdom, and the United States.

No one could say how long the Russians would cooperate with the committee they had been instrumental in setting up.

A Compromise Space Program at Home

Moved as they were by the horrors of a possible space war, the President and his foreign policy advisers clung tenaciously to the space-for-peace policy. The Chief Executive did not compromise this position until after Sputnik when he regretfully conceded the need for a military space program--but one of small dimensions. He still hoped to focus world attention on America's interest in peace by emphasizing the civilian-scientific program for the exploration of space. Sputnik compelled a compromised space policy at home, but the extent of the compromise was made clear only in the chronology of events.

The statement of 9 October 1957 that the United States would not engage in a space race was not reassuring to the military. Then another month passed before there was any indication from the White House what the national policy would be. Sputnik II on 3 November occasioned another outcry of protest from the press, and four days later, President Eisenhower addressed the nation by television. His intent was admittedly to reassure the uneasy public on the advanced status of American weapons, particularly missiles, and he announced the appointment of Dr. James R. Killian, president of the Massachusetts Institute of Technology, to be Special Assistant to the President for Science and Technology. Killian would be aided by the Science Advisory Committee.[20]

The appointment of Killian at this time appeared to recognize the inevitability of a military space program. The President confirmed this idea on 5 February 1958 in a press conference when he said the Department of Defense would "continue" to control military space projects even after the establishment of a civilian space agency.[21] At the same time, the President's Science Advisory Committee was working on the first comprehensive statement of U.S. interests in space. Simultaneously, the Department of Defense assumed that a military program was certain and planned accordingly.

On 26 March 1958 the advisory committee released a policy paper. It listed three reasons why space should be explored: to acquire scientific knowledge, further national prestige, and guarantee American military strength. This was the first top-level indication of what policy would prevail. On 2 April the President committed himself officially. He asked Congress for a civilian agency, the National Aeronautics and Space Administration (NASA), to conduct all space activities except those primarily associated with military requirements. The Presidential message was confirmation of a two-fold space program--one civilian and one military.[22]

Congress debated the nature of NASA for several months and did not pass the National Aeronautics and Space Act of 1958 until midsummer. The President approved it, PL 85-568, on 29 July, and the dual space program became statutory.

Meanwhile, the Secretary of Defense initiated a move to make the military more sure of their space responsibilities. In March 1958, he suggested that NSC's Planning Board consider the advisability of NSC issuing a national security policy on space. At once the board set up the Ad Hoc Subcommittee on Space, which requested and received comments and assistance from the National Science Foundation, the Central Intelligence Agency, the three services, and other agencies throughout the Government. The product was the Preliminary U.S. Policy on Space, more conveniently known as NSC 5814/1, which the President approved on 18 August 1958.[23]

The document set forth more explicitly than hitherto the purpose and principles of the civilian and military programs. NSC recognized that space had military significance but was more concerned with the political implications. It was politically dangerous for Russia to remain permanently superior to the United States in astronautics, and the penetration of space made it more necessary than ever to work toward international control and cooperation. In conclusion, the council advocated a six-point policy: continue the IGY experiments; recognize the interest of the United Nations in space; propose a series of bilateral arrangements with other nations, including Russia, to regulate current activities in space; invite all nations to participate on a reciprocal basis in U.S. scientific projects; propose other projects for multilateral participation; and assist other nations of the Free World in their space projects.[24]

It is important to note the council's insistence upon international cooperation in space, with the emphasis on reciprocal development of space science by the United States and other nations. All American contributions to these activities would come within the responsibility of NASA, and, in a very limited but very real sense indeed, NASA was certain to be an adjunct of the Department of State.

By the late summer of 1958, there were three documents that, taken together, expressed the Administration's space policy--the report of the President's Science Advisory Committee on 26 March, the National Aeronautics and Space Act of 29 July, and NSC's Preliminary U.S. Policy on Space of 18 August. Each affirmed that there must be a military space program, but the overall tone was that the military program should be kept as small as possible.

A fourth important document appeared in March 1959--the first Operations Plan of NSC's Operations Coordinating Board (OCB). The paper was intended to guide and implement the national program. It recommended a four-point action to include analysis, incident by incident, of international legal issues as they developed; negotiation of international agreements for a complete record of satellite orbits and frequencies; formulation of agreements with other nations as required for the peaceful use of space; and preparation of world opinion psychologically and politically for the possible launching of American reconnaissance satellites. In its general approach, the OCB Operations Plan indicated a slight change of thinking, at least within the confines of NSC, that meant modification of the space-for-peace policy along lines a little more favorable to the military.[25]

By the time the OCB plan was finished, there had been many important changes in the situation since August 1958 when the President had approved the preliminary policy statement. Both the military and scientific space programs had gained significant new data on the space environment, the organization of the civilian and military space programs had been completed, and the international situation demanded more than ever that the United States regain its lost prestige. Under the circumstances, the NSC felt that the policy statement required a "complete review" and, with the President's approval, entrusted the work to an ad hoc committee of the National Aeronautics and Space Council (NASC), which the Space Act of 1958 had brought into existence.[26]

The ad hoc committee began its work in July, using as reference the earlier-noted executive and legislative measures and then, in August 1959, NSC's National Security Policy or NSC 5906/1. This policy statement affirmed the necessity from a national security viewpoint of a space program that could support the scientific, military, and political aims of the United States.[27]

In November 1959, NASC transmitted the report of its ad hoc committee to NSC. The report urged a national space policy that would:

> Carry out energetically a program for the exploration and use of outerspace by the United States, based upon a sound scientific and technological progress, designed: (a) to achieve that enhancement of scientific knowledge, military strength, economic capabilities, and political position which may be derived through the advantageous application of space technology and through appropriate international cooperation in related matters, and (b) to obtain the advantages which come from successful achievements in space.

In addition, the report declared that civil, scientific, and military space projects had important implications for national security, and it regretted that the Soviet's spectacular "firsts"--which by then included the orbiting of a Sputnik with canine passenger, an interplanetary probe, and a lunar impact--had raised Russian international prestige even above the level attained in October 1957. Though the full military significance of space could not then be defined, it was apparent that space vehicles would have to be employed to enforce whatever international agreements might eventually be reached to prevent a space war, and until then reconnaissance satellites could be a safeguard against another Pearl Harbor strategy.[28]

The recommendations served NASC and NSC in their task of revising the policy on space. The paper, NSC 5918/1, was completed 12 January 1960 and signed by the President on 26 January. It represented no great change from all that had come before. It admitted the importance of space but kept the emphasis on the civilian program. The Administration remained consistent in downgrading the military space effort from March 1955, with the initiation of Vanguard, through the first reactions to Sputnik, and across the months of 1958 and 1959. The most important change had come in March 1958 when the President's Science Advisory Committee admitted need for a military program, but after that the emphasis tarried on the same low plateau.

There was, however, a marked difference between expressions of national policy and the actual implementation of that policy. Once a military space program became permissable, it gained a momentum from its own projects that did not completely respond to the brakes of policy. As a result, the status of the military program was far more advanced in the summer and autumn of 1959 than the words of the National Security Council papers could indicate.

The success of the military projects was all the more remarkable because the program, as a whole, became entangled in the web of organizing inside and outside the Department of Defense.

VI. ORGANIZING FOR THE MILITARY SPACE PROGRAM

The month between Sputnik I and Sputnik II, 4 October to 3 November 1957, was filled with criticism of the American missile-space programs. The two were almost identical in the public mind, and in truth, space projects were and would be for some time almost completely dependent upon missile organizations and components. It was a critical hour in the nation's history, and the demand for action was not to be ignored. Time and again the question was raised as to why the missile-space programs had failed to meet the crisis and how the programs could be vitalized to carry forward the burden of catching up with Russia.

There was a tendency among some to attribute the failure to inter-service rivalry or "service bickerings." The President on more than one occasion publicly declared that interservice rivalry must stop, and his comment was so placed in context as to imply that this was the evil of the day.* Some periodicals took up the cry to designate a "Pentagon boss" capable of ending "service bickerings" and put the nation ahead of the Soviet Union in technology.[1]

A contrary viewpoint held that the failure to win first place in space was not due to "service bickerings" but to national policy. There was no vigorous American space program in 1957 because of a preference

---

*In his television address to the nation on 7 November 1957 the President said, "Inter-service competition shall not be allowed to /harm/ . . . our scientific and development program." Again, in his State of the Union message to Congress on 9 January 1958 he said: "I am not attempting today to pass judgment on the charge of harmful service rivalries. But one thing is sure. Whatever they are, America wants them stopped."

for economy, an insistence that space projects must offer returns commensurate with cost, and a determination to keep the military out of space for the sake of foreign relations. The decisions were made over a period of 12 years, 1945 to 1957, by the Research and Development Board, the Department of Defense, the Department of State, and the White House. National policy said "no" both to Navy and Air Force space ambitions in 1946-48, rejected Project Orbiter in 1955, dulled the Vanguard effort for two critical years, and refused permission for ABMA to launch a satellite.

There was also widespread objection to the appointment of any more bosses. In 1957 many voices cried out in Congress, in the armed forces, and among interested citizens to simplify rather than elaborate the missile organizational setup to which the space program was certain to be tied for some time. The need was not for more "czars" with overlapping domains and authorities, but for right decisions.[2] Missile organization charts showed "bureau on top of bureau, committee on top of committee, office on top of office . . . . /To/ the average unsophisticated, or even sophisticated person, it looked like the most complicated jigsaw puzzle that ever was invented." The question then, in 1957, was whether the organization was to be more simple or more complex, whether the czars were to be overthrown or perpetuated.[3]

Serious efforts were made to escape from the labyrinth. On 17 October, General Putt, DCS/Development at Headquarters USAF, acting on orders from higher authority, directed Lt. Gen. Samuel E. Anderson, the commander of ARDC, to assemble an ad hoc committee to consider ways by which the Air Force could assist in countering world reactions to Sputnik I. The committee was composed of members of the Air Force Scientific Advisory Board (SAB) and the aircraft industry, plus a small group of ARDC personnel as technical advisers. The committee met 21-22 October under the chairmanship of Dr. Edward Teller. The two-day discussion produced an impressive report that exhibited no shyness of the truth. In the technological war between the United States and the Soviet Union, the former had slipped behind because of complacency and swollen bureaucracy. "To date, our administrative and management practices have not permitted either the responsible civilian or Armed Service agencies to establish a stable yet imaginative R&D program." The committee's two recommendations were strongly phrased:[4]

> 1. Consolidate the organization and simplify the management for the development and operation of ballistic missile and space flight programs from the Office of the Secretary of Defense on down, including the efforts of all services.
>
> 2. Put the ballistic missile and space flight programs on a maximum effort basis in all its aspects, without reservation as to time, dollars, or people used. Most important of all, provide a realistic assurance that the entire program has the priority of governmental and national interest required by the threat.

The Teller Report, which bore the signatures of some very distinguished scientists and leading authorities* on missiles and satellites, was circulated on 28 October among high levels of the Department of Defense. By coincidence, Trevor Gardner's article, "But We Are Still Lagging," appeared in Life one week later, on 4 November. Gardner, too, argued for a simplified organizational arrangement to meet missile-space program requirements and for ample funds to support research and development. Since Gardner's

---

*List of members of the Teller Committee: Edward Teller, E.J. Barlow, J. Beerer, K.J. Bossart, G.H. Clement, E.B. Doll, W.R. Dornberger, K. Ehricke, C. Faulders, C.L. Forrest, D.T. Griggs, M.D. Hunter, J. Isenberger, T.G. Lanphier, F. O'Green, W.F. Parker, L.D. Ridenour, R.J. Sandstrom, M. Sherman, W.M. Sith, E. Spraitz, E.A. Steinhoff, G.S. Trimble, G.E. Valley, T.F. Walkowicz, R.H. Widmer, R.G. Wilson. Also attending were representatives of WADC, AFCRC, AFBMD, and AFOSR.

forebodings had already been tragically justified, he was not a person whose advice should have been ignored. On 7 November, within three days of Gardner's article, the Security Resources Panel of ODM, under the chairmanship of H. Rowan Gaither,* submitted its report, Deterrence and Survival in the Nuclear Age, which went to NSC and therefore came within the President's advisory circles. Along with the Rockefeller report, International Security--the Military Aspect, which appeared in January 1958, the Gaither report was part of a rising tide of criticism of the Government's overly complex and inadequately budgeted programs.

In the midst of this criticism and debate the President announced on 7 November his selection of Dr. Killian as his scientific adviser. Insofar as his appointment indicated some form of a military space program, the services were pleased. Insofar as the appointment might indicate making the missile-space program more and more complex, the military were uneasy. At the same time the Secretary of Defense, acting on the assumption that there would be a military space program, showed clearly that he too was thinking of adding to the number of military missile-space agencies. By the end of 1957 it was evident that the age of the czars had not passed.

Advanced Research Projects Agency

On 7 August 1957, President Eisenhower announced the resignation of Wilson and the nomination of Neil H. McElroy as Secretary of Defense. Toward the end of September, McElroy came to the Pentagon to familiarize himself with the job he would occupy on 9 October. Thereafter he visited some of the major military installations, and on 4-5 October he was guest of the

---

*Gaither was former president of the Ford Foundation. His committee began its study in April 1957, but before it was completed he became ill and was succeeded by two co-chairmen--Robert C. Sprague and William C. Foster.

Army Ballistic Missile Agency in Huntsville, Ala. Along with Maj. Gen. John B. Medaris, commanding general of ABMA, and Von Braun, McElroy was among the first to learn of Sputnik I. No one could have had a more dramatic induction to high office. (See testimony, p 104.)

Almost immediately McElroy found himself in the midst of a reorganization that could not always be clearly understood. Obedient to Presidential direction, the Secretary of Defense abolished his Office of Special Assistant for Guided Missiles and created in its place the Office of Director of Guided Missiles to "direct all activities in the Department of Defense relating to research, development, engineering, production, and procurement of guided missiles." William Holaday, who had been the Special Assistant for Guided Missiles, headed the new office, apparently clothed, at the President's behest, with the authority of the Secretary of Defense in the field of guided missiles. Presumably his duty was to override service rivalries. At once, however, the Secretary of Defense said that Holaday could not direct the work of the services in the field of guided missiles, and there were some questions on Capitol Hill on how Holaday could be a director if he could not direct. Holaday himself was vague about his authority and did not know what his relationship was to Dr. Killian, the President's Special Assistant for Science and Technology. McElroy stated that he too was uncertain what authority Dr. Killian possessed.[5]

Though the post-Sputnik domains of authority were thus far from sharply drawn, McElroy proceeded to plan for yet another czar within DOD whose duty would be to unify the space projects scattered among the three services. He first spoke of this newly conceived "special projects" agency

Testimony of Neil H. McElroy, SOD,
27 November 1957

Senator Symington: I know that you came into the Department [of Defense, Mr. McElroy, just about the time that the Russians launched Sputnik] . . . . Were you surprised when they launched the sputnik?

Secretary McElroy: I was very much surprised. In fact, I was down at Huntsville, Ala., having just spent the day examining Jupiters, and I am unlikely to forget the time that I heard about the first Sputnik. It certainly launched me into a job here on certain wings. So that would be clear to me as long as I live.

Senator Symington: Do you remember that some people did not seem to be particularly surprised?

Secretary McElroy: I do, and I suppose if I had been privy to the intelligence knowledge that had been around in the community, I would not have been so surprised, either. But I was very much surprised.

Senator Symington: Well, I was surprised some people were not surprised; because when defense authorities came before our Subcommittee on Appropriations last August, and asked for money, they said--I want to be sure I state it correctly--that this money was asked for so we would "launch the first artificial satellite."

(Senate Hearings before the Preparedness Investigating Subcmte, 85th Cong, 1st & 2d Sess, Inquiry into Satellite and Missile Programs, p 250.)

when he appeared before a congressional committee on 20 November 1957. He said then that he would pace the agency at a level above that of the three services so that it could control interservice rivalry. The director would then be responsible for all military research and development efforts "in the satellite and space research field" and for antiballistic missiles.[6]

When McElroy sought the opinion of the Joint Chiefs of Staff, he found opposition. The service chiefs did not want the agency to have development and contractual authority, because they felt such an arrangement would hamper the transition of systems from development to operational status. The need was for an office with authority to make policy decisions. The services were quite capable of managing their research and development if they could but be authorized to proceed with the work. The Joint Chiefs submitted these views to the Secretary on 25 November 1957.[7]

McElroy overruled the JCS objections and continued his plans for the Advanced Research Projects Agency (ARPA), as it had then come to be called. The Secretary received full support from the President, who asked Congress to give ARPA a budget but one that would be largely spent through the technical and procurement agencies of the Army, Navy, and Air Force.[8]

McElroy wanted to act as quickly as possible. The National Security Act amendments of 1949 had vested the Secretary of Defense with authority to transfer, reassign, abolish, or consolidate noncombatant functions after notifying Congress. On the advice of his General Counsel, McElroy assumed he thereby had the authority to establish ARPA, but the Senate and House did not agree to this interpretation. Without any desire to hinder the Secretary, they questioned his right to set up ARPA.[9] In order to assist McElroy, Congress included the following provision in Public Law 85-325, which the President approved on 12 February 1958:

> The Secretary of Defense or his designee is authorized to engage in such advanced projects essential to the Defense Department's responsibilities in the field of basic and applied research and development which pertain to weapons systems and military requirements as the Secretary of Defense may determine after consultation with the Joint Chiefs of Staff; and for a period of one year from the effective date of this Act, the Secretary of Defense or his designee is further authorized to engage in such advanced space projects as may be designated by the President.

The purpose of authorizing the Secretary to engage in advanced space projects for one year was to insure the continuation of Project Vanguard and other "peaceful" space ventures that might emerge while Congress decided on the nature and organization of the national space program. On 11 February, Congress also passed Public Law 85-322 to provide for the transfer of $10 million from the military services to ARPA, thus insuring the agency an independent budget.

Completely confident that Congress would empower him to activate ARPA, McElroy had established the agency on 7 February. At the same time he gave it a broad charter, with authority to direct such research and development projects being performed within the Department of Defense as the Secretary might assign.[*] The charter further authorized ARPA to arrange for the performance of the work by other governmental agencies, including the three services. It was also possible for ARPA to contract with individuals or institutions and acquire test facilities and equipment as approved by the Secretary of Defense.[10]

Roy W. Johnson became ARPA's first director. He had first to

---

[*]Somewhat later McElroy gave ARPA specific responsibilities for research and development activity on ballistic missile defense, propellant chemistry, and military space.

delineate the authority areas of his office and of the offices of the Assistant Secretary of Defense (Research and Engineering) and the DOD Director of Guided Missiles, under Paul D. Foote and Holaday, respectively. Johnson, Foote, and Holaday recognized that the relationship of their agencies had to be one of close interdependence to permit a constant exchange of information in their respective fields. However, the relationship between ARPA and the Director of Guided Missiles was made closer by the fact that many of the vehicles and components employed by guided missiles and space vehicles were identical. On the other hand, both agencies would be dependent for further progress on the products in the broader fields of research under the authority of the Assistant Secretary (R&E).[11]

Johnson organized ARPA in three divisions--Financial Management, Policy and Programs, and Technical Operations. He obtained a large part of his staff by a contractual arrangement with the Institute for Defense Analysis (IDA),* which provided a unit of 40 persons headed by Dr. Herbert F. York. The latter was already well known for his thermonuclear work at Livermore Laboratory, and in ARPA he served as Johnson's chief scientist. By May 1958, ARPA was an operating organization, and its chief weakness was the lack of experience on the part of IDA personnel with military methods of procedure.

Director of Defense Research and Engineering

At the same time that plans were being made for ARPA, the President and the Secretary of Defense were preparing to reorganize the Department

---

*IDA was created in 1955 in contract between OSD and Massachusetts Institute of Technology to supply qualified personnel for the Weapons Systems Evaluation Group. MIT initiated the work and invited five other universities to participate. Ford Foundation granted $500,000 for working capital.

of Defense in a way certain to influence the space program. In his State of the Union message of 9 January 1958, Eisenhower, referring to interservice rivalry, said some weapons did not fit into any existing service pattern and gave rise to "jurisdictional dispute." He felt that the situation demanded important changes in the organization of the Department of Defense and stated he would later send specific recommendations to Congress.[12]

Three months afterwards, on 3 April, the President submitted his request. He said that "separate ground, sea and air warfare is gone forever" and that peacetime activity of the military forces should be completely unified. He wanted the authority of the Secretary of Defense to be "clear and direct" in respect to the development of new weapons. Therefore, one of his important points was the elimination of the Office of Assistant Secretary of Defense (Research and Engineering) and in its place the establishment of a Director of Defense Research and Engineering (DDR&E), with three major functions:[13]

> first, to be the principal advisor to the Secretary of Defense on scientific and technical matters; second, to supervise all research and engineering activities in the Department of Defense, including those of the Advanced Research Projects Agency and of the Office of the Director of Guided Missiles; and third, to direct research and engineering activities that require centralized management.

The President apparently intended the Director of Defense Research and Engineering to outrank the ARPA director as well as the Director of Guided Missiles.

After due deliberation Congress enacted Public Law 85-599, Department of Defense Reorganization Act of 1958. Among its provisions was the establishment of a Director of Defense Research and Engineering to be appointed by the President and taking precedence within the Department of

Defense after the Secretary and Deputy Secretary of Defense and the three service secretaries. He would be the principal adviser to SOD on scientific and technical matters; supervise all research and engineering activities in DOD; and direct, control, assign, or reassign the research and engineering activities deemed by the Secretary of Defense to require centralized management. The President approved the act on 6 August 1958 and on 24 December appointed Dr. York as the first director of the new agency.[14]

Activation of Directorate of Advanced Technology

Prior to Sputnik I Air Force space activities had been handled in the Office of the Deputy Chief of Staff/Development, with Brig. Gen. Homer A. Boushey, Deputy Director of Research and Development, responsible for the overall coordination of those projects that pertained to space.

On 22 November 1957, two days after McElroy publicly spoke of his plans for ARPA in congressional hearings, Col. V.Y. Adduci, Assistant Director of the Office of Legislative Liaison, urged the Air Force "to jump the gun on the problem of astronautics by appointing either a Director or Assistant Chief of Staff for Astronautics." In view of the growing opposition within Headquarters USAF to the creation of additional assistant chiefs of staff, there was little probability of placing an agency at that level. Conceivably, it could have been located in the Office of the Assistant Chief of Staff for Guided Missiles, where there was already some capability for the work. The Chief of Staff decided, however, to place the agency under the DCS/Development, and on 10 December, General Putt announced the establishment of the Directorate of Astronautics, to be headed by General Boushey.[15]

OSD reacted unfavorably. Holaday publicly stated that the Air Force "wanted to grab the limelight and establish a position." The Secretary of Defense expressed his opposition to the use of the term "astronautics," which seemed to him an Air Force bid for popular support. Strong pressure on Headquarters USAF from above, verbal rather than written, made it advisable on 13 December for Putt to cancel his directive of 10 December.[16]

Headquarters USAF, keenly aware of the need to centralize its space activities in some one agency, regarded the cancellation of 13 December as merely a postponement. The prospects of getting OSD approval, however, was admittedly slight for the next few months, and an interim measure was needed. Since space vehicles were dependent on ballistic missiles, Headquarters adopted the temporary solution on 4 March 1958 of authorizing the Assistant Chief of Staff for Guided Missiles to coordinate USAF space activities.[17]

At about the same time the DCS/Development suggested the advisability of requesting OSD approval of an Air Force space agency. The Office of the Chief of Staff was not averse but foresaw a long delay. There were weeks of negotiation between the Air Force, OSD, and ARPA. An Air Force space directorate, it was argued by the Air Staff, was needed to serve as liaison with ARPA, and it would be equally needed as a means of contact with the civilian space agency then being provided by Congress. Plans were carefully drawn, and on 22 July, after Congress had passed the space act, Secretary Douglas formally requested permission to activate the directorate. Two days later Deputy Secretary of Defense Quarles gave his approval. Even then, the term "astronautics" was considered impolitic for the military. On 29 July, General White issued General Order 44, stating that "the verbal order of the Chief of Staff establishing the Directorate of Advanced Technology, Deputy Chief of Staff, Development, effective 15 July 1958, is confirmed."[18]

There was no directorate charter at the time, but the DCS/Development summarized the purpose of the agency in a 29 July memorandum:[19]

> To supervise at the Air Staff level the formulation of the Air Force Advanced Technological Program; provide technical information and advice to the Air Staff on the process of developments; maintain coordination with ARPA, the Departments of the Army and Navy and other interested government agencies; and maintain liaison with civilian educational institutions, industry, and representatives of foreign governments engaged in research and development activities.

The same memorandum named General Boushey as director and provided him with a small staff. Boushey promptly organized his directorate under four assistants--for Boost Glide Systems, Space Projects and Systems Studies, Manned Military Space Systems, and Unmanned Military Space Systems.

Doubtless Headquarters USAF hoped to make the Directorate of Advanced Technology the control point for all Air Force space projects. However, since the space projects were dependent upon missiles, the space program necessarily involved AFBMD, whose main point of contact with Headquarters was through the Assistant Chief of Staff for Guided Missiles. Under the circumstances it was imprudent to sever all ties between the guided missile office and the space program, and a reassignment of authority between the Assistant Chief of Staff for Guided Missiles and the Directorate of Advanced Technology was inevitable. On 6 April 1959 the Chief of Staff rescinded the directive of 4 March 1958 and delegated responsibility for coordinating and monitoring all Air Force space activities within the Air Staff to the Directorate of Advanced Technology. However, the Assistant Chief of Staff for Guided Missiles continued to retain responsibility for coordinating the requirements for ballistic missile resources needed in support of the space projects, including boosters and test facilities. On 13 April a Headquarters office instruction defined the relationship between the Directorate of Advanced Technology and other offices of the Air Staff, ARPA, and NASA.[20]

VII. CIVILIAN SPACE AGENCIES

Both houses of Congress were deeply disturbed by Sputnik I and Sputnik II. The Russians appeared well on the way toward an ICBM-atomic-war capability that would permit direct attack on American cities and industry. Equally disconcerting, from the viewpoint of the cold war strategy, was the detrimental impact the Russian successes undeniably had on the prestige of the United States. Moreover, the United States was far behind the Soviets in planning and conducting space activities--an important factor in such areas as international law, foreign relations, and hitherto unimagined weaponry of offense and defense, as well as a compelling appeal to imagination through projects of such universal usefulness as meteorology and navigation. In the yet vaster areas of pure science, space operations seemed destined to be of incalculable importance to the whole human race and its social structure.

Senators and representatives did not content themselves with expressions of astonishment, dismay, or incurable optimism. The situation was serious, and Congress prepared for serious action. On 27 November 1957 the Preparedness Investigating Subcommittee of the Senate Committee on Armed Services opened an investigation of the American missile and space programs. On 6 February 1958 the Senate established a Special Committee on Space and Astronautics. The House followed suit, establishing on 5 March its own Select Committee on Astronautics and Space Exploration. Meanwhile, both the Senate and House came to the assistance of the Department of Defense by cooperating with Secretary McElroy in establishing

ARPA. This DOD agency assured the nation that there would be a military space program. No one could say at the time how much the program would be curtailed by the civilian-scientific program still being debated in White House and congressional circles.¹

By March 1958, Congress was conversant with several alternative proposals for the organization of space agencies by the executive branch, as well as several ways in which the legislative branch could keep itself informed. The congressional committees felt that the President had a wide choice. He could entrust the entire space program to one of the following: the Department of Defense, the Atomic Energy Commission, a new commission modeled on AEC, a department of science, or a coordinated effort by the National Academy of Science and the National Science Foundation. In exercising its own watchdog prerogatives, Congress could choose between creating a new joint committee on space, adding space responsibilities to the Joint Committee on Atomic Energy, or placing the existent Senate and House space committees on a permanent basis.²

Congress had not gone beyond this point when the President's Science Advisory Committee issued its statement of 26 March 1958.* The paper showed that the military space program was certain to be continued. It showed also that an extensive civilian-scientific program would be undertaken and that Congress would be called upon to establish by law a civilian space agency, or a complex of space agencies. The situation then moved rapidly toward its climax.

Hearings on the President's Proposed Space Agency

On 2 April 1958 the President forwarded to Congress his recommendation "that aeronautical and space science activities sponsored by the United

---

*See above, p 94.

States be conducted under the direction of a civilian agency, except for those projects primarily associated with military requirements." He urged Congress to create a National Aeronautics and Space Agency into which the National Advisory Committee for Aeronautics (NACA) would be absorbed as a nucleus. In this way NASA would continue the aeronautical research functions of NACA and expand into the space area. The new agency would be headed by a director appointed by the President with the consent of the Senate. President Eisenhower requested the creation also of a national aeronautics and space board to advise him, with representation from interested government agencies including the military. He added:[3]

> It is contemplated that the Department of Defense will continue to be responsible for space activities peculiar to or primarily associated with military weapons systems or military operations. Responsibility for other programs is to be assumed by the new agency.

The President clearly called for a space program that was split between civilian-scientific interests and the military. It was then up to Congress to approve the President's policy in such a way as to insure the security of the nation in space. This point, indeed, became the crux of the long and interesting hearings conducted by the Senate and House space committees. Congress could not forget that space exploration was possible in 1958 because of missile developments. Also, in 1958 the military controlled most of the personnel trained for research in space propulsion and vehicles as well as the materials needed for the future program. For Congress, the most obvious and immediate problem was to determine as exactly as possible the relationship between the civilian and military programs.[4]

During April and May a procession of distinguished witnesses moved before the congressional space committees. At first the consensus was that NACA, with a few changes in its charter, would become NASA. The scientists sanctioned such an arrangement. From their point of view a civilian-scientific program was essential because the nonmilitary aspects of space exploration were too important to be entrusted to a purely military program. Only a civilian-scientific program could insure a technically sound approach. Yet the scientists were also of one accord that military interests should be safeguarded, and they spoke specifically of reconnaissance and communication satellites. These witnesses were confident that military applications would follow automatically from a scientific program.[5]

The military and their representatives were in general agreement with the civilian scientists, but they interjected a few cautious reservations. Spokesmen for the Department of Defense approved the establishment of NASA and spoke of it as being an extension of NACA into space. However, all of them spoke out against excluding the Department of Defense from basic research for service missions. This precaution would entail avoidance of a rigid definition of weapon programs. Conceding that the nonmilitary aspects of the national space program should be under civilian direction, the point was made time and again that nothing should be done to prevent the Department of Defense from anticipating "reasonable requirements" and proceeding with the work immediately. The military theme was simply that there should be two programs and they should be closely coordinated.[6]

The National Aeronautics and Space Act of 1958

Congress enacted on 16 July the National Aeronautics and Space Act of 1958. In an introductory declaration of policy and purpose, Congress affirmed that the space activities of the United States were devoted to peaceful ends and that responsibility for conducting this work was vested in a civilian agency. The authority of the agency was then qualified by important exceptions. Activities primarily associated with weapon system developments, military operations, or the defense of the United States--including the necessary research and development--"shall be the responsibility of, and shall be directed by, the Department of Defense." The act authorized the President to determine which agency, civilian or military, should have responsibility for specific projects.[7]

The Space Act provided for three new agencies. Two of them were wholly civilian--the National Aeronautics and Space Council (NASC) and the National Aeronautics and Space Administration (NASA). The third agency, the Civilian-Military Liaison Committee (C-MLC), was, as its name implied, hybrid. The respective purpose of these agencies was to assist and advise the President in space matters, to direct the civilian-scientific space program, and to tie together the civilian and military program in "a two-way street of information and decision making."[8]

The council consisted of the President, Secretary of State, Secretary of Defense, Administrator of NASA, Chairman of AEC, and four additional members appointed by the President--one from within and three from outside the Government. NASC would assist the President to survey aeronautical and space activities and "provide for effective cooperation between the National Aeronautics and Space Administration and the Department of Defense."[9]

NASA, headed by a presidentially appointed administrator, received authority to plan, direct, and conduct aeronautical and space activities; arrange for participation in space activities by the scientific community;

and provide for the widest practicable dissemination of acquired information. NASA was thus unmistakably an operational agency and would require operating facilities as soon as activated. The need was met by absorbing NACA, its personnel, and facilities. The act directed all other government departments and agencies to cooperate as required by NASA "in making their services, equipment, personnel, and facilities available." The act also stated that NASA, under the guidance of the President, could engage in a program of international cooperation, a provision that gave a foreign policy tie-in with the Department of State.[10]

The Civilian-Military Liaison Committee would consist of a chairman appointed by the President and a membership of unspecified number but equally divided between representatives from NASA and DOD. The military representatives in turn would be equally divided between OSD and each of the three services. "The Administration /NASA/ and the Department of Defense, through the Liaison Committee, shall advise and consult with each other on all matters within their respective jurisdictions relating to aeronautical and space activities and shall keep each other fully and currently informed with respect to such activities." In case of unresolved disagreement between the Administrator of NASA and the Secretary of Defense, either of them could refer the matter to the President for decision.[11]

The train of witnesses from the Department of Defense had ably impressed on Congress the necessity of conducting research and development for its own space projects. Congress in turn went to some length to insure DOD's freedom in this field, as explained in the Conference Report:[12]

> The Congress recognizes that the development of aeronautics and space capabilities is important both to peaceful purposes and to the defense of the United States and for the preservation of peace everywhere. It is the intent of Congress that the necessary freedom to

carry on research, development, and exploration be afforded both a civilian agency and the Defense Establishment to insure the full development of these peaceful and defense uses without unnecessary delay, to exclude the possibility that one agency would be able to preempt a field of activity so as to preclude the other agency from moving along related lines of development necessary to the full accomplishment of its duties assigned under this act. At the same time, such freedom to pursue activities should be so conducted as to avoid unnecessary duplication of effort and expenditure. This can be accomplished by providing for full cooperation between the civilian agency and the Department of Defense. It is clearly recognized that activities which are peculiar to or primarily associated with weapons systems or military operations or to the defense of the United States (including the research and development necessary to make effective provision for the defense of the United States) shall be under the jurisdiction of the Department of Defense. However, because there is a gray area between civilian and military interests, and unavoidable overlapping, it is necessary that machinery be provided at the highest level of Government to make determinations of responsibility and jurisdiction.

This act makes such provision by providing that the President, assisted by an Advisory Council, shall make the actual determinations in the assignment of new programs and projects. The act also provides that the Administrator of the National Aeronautics and Space Administration and the Secretary of Defense can seek solutions to questions of jurisdiction either directly or through a Civilian-Military Liaison Committee to hold to a minimum the questions referred to the President and the Council.

Organizing Space Agencies under PL 85-568

The President approved the Space Act on 29 July 1958. Under its terms it would become effective on a convenient date within the succeeding 90 days. This allowed the President and his advisers 13 weeks, until 26 October, in which to appoint the members of the space council and the top NASA officials. Since C-MLC was a liaison committee between the Department of Defense and NASA, the appointment of its chairman could await the actual activation of NASA.

On 8 August 1958 the President selected Dr. T. Keith Glennan, president of Case Institute of Technology, Cleveland, Ohio, and Dr. Hugh L. Dryden, Director of NACA, as the administrator and deputy administrator of NASA. On 4 September the President chose William A.M. Burden as the fifth government member of the space council. At the same time the President appointed Drs. James H. Doolittle, Alan T. Waterman, and Detlev W. Bronk as the nongovernment members. On 31 October the President reassigned William Holaday from Director of Guided Missiles to chairmanship of C-MLC.

The Space Act did not specify how NASA was to be organized other than to provide for an administrator and a deputy administrator. Glennan therefore had a free hand in setting up the agency, and he acted with dispatch. He organized NASA into three divisions--Space and Flight Development, Aeronautical and Space Research, and Business Administration. On 1 October 1958, Glennan announced that NASA was prepared to discharge its duties.[13]

Of the three agencies established by the Space Act, the Civilian-Military Liaison Committee was the least well defined, and it became the most difficult to organize. Since Congress did not fix the membership, it was up to Glennan and McElroy to make the arrangement. After several conferences there was an agreement that the committee would be composed of four representatives from NASA, one from OSD (ARPA), and one each from the Army, Navy, and Air Force. These eight, along with the chairman, gave the committee nine members.[14]

On 12 September, McElroy asked the three services to recommend their C-MLC representatives and alternates. The Air Force had already given much thought to this, being deeply concerned by the fact that NASA would absorb NACA along with much of the space program originally conceived by the Air Force and still considered essential to its mission. Thus some Air Staff officials felt that an Air Force general officer should be chairman

of C-MLC since the Air Force had a predominant role in both aeronautics and space. The suggestion was not received enthusiastically, and on 31 October, with Holaday's selection as chairman, the Department of Defense announced its committee representatives.*[15]

A much more difficult question concerned the scope of C-MLC's functions. The Space Act had been vague on this point, and it was generally agreed that a charter or similar paper was necessary. The Joint Chiefs were uneasy least the military be unable to convey their viewpoint to the civilian agency, and therein they doubtless reflected the anxiety of the three services as well.[16] Negotiations begun in September resulted in a first draft circulated among the services during the first week in October. The Army, Navy, and Air Force all wanted a more convincing guarantee of future cooperation between NASA and DOD. In addition, the Air Force still argued for a USAF general officer as chairman.[17]

A series of high-level conferences ensued involving the service secretaries and the Director of ARPA. Out of these conferences came a compromise draft that reconciled the NASA-DOD viewpoints. The charter stated that C-MLC would provide a channel for the exchange of information and advice between NASA and DOD, encourage further NASA-DOD contact at appropriate levels, recommend courses of action in the event of differences between NASA and DOD, and perform other duties as assigned by NASA or DOD. The committee would meet once each month and report its conclusions to NASA and DOD.[18]

---

*The military representatives were Roy W. Johnson, OSD; Maj. Gen. W.W. Dick, Army; Vice Adm. R.B. Pirie, Navy; and Maj. Gen. R.P. Swofford, Air Force. The NASA representatives, announced on 17 November 1958, were Hugh L. Dryden, Abe Silverstein, Homer J. Stewart, and Ira H. Abbott.

The Armed Forces Policy Council (AFPC) approved these terms of reference on 22 October, and this action, in the opinion of the Director of ARPA, was equivalent to ratification by DOD.[19]

The AFPC approval of the C-MLC charter, and even the appointment of Holaday as chairman, did not completely clarify the position of the committee in the overall structure of space organizations. It was impossible to predict how the responsibilities of the committee would develop, and during the formative period the Air Force needed a particularly sensitive channel of contact to permit prompt action. Lt. Gen. Roscoe C. Wilson, DCS/Development since March 1958, appointed a member of his staff to monitor C-MLC activities for the USAF member of the committee. He, in turn, was supported by a designated officer from each of three directorates of DCS/Development: Advanced Technology, Requirements, and Research and Development.[20]

Though much thought went into the organization of C-MLC and into the selection of its members, between November 1958 and July 1959 the agency functioned only as an atrophy. So unimportant were its contributions to the space program that its history can be largely ignored.*

Organization for Space at End of 1958

In October 1957 the cry had been for a simplification of the missile-space complex within the Department of Defense. A year later the missile

---

*The first C-MLC charter was a compromise and did not allow the committee the scope of activity undoubtedly intended by the Space Act. From the beginning it was largely ignored both by DOD and NASA, with most of the important issues being settled directly by the Secretary of Defense and the Administrator, NASA. As a consequence the minutes of C-MLC's first eight meetings, beginning 25 November 1958 and continuing through 18 June 1959, exhibit a poverty of activity, and even after the beginning of fiscal year 1960, when a new charter went into effect, there was little improvement.

complex had not been simplified, the military space complex had been elaborated, and in addition there was the newly created civilian complex. Since the military were obligated to provide NASA with much of its logistic support, it was not always a simple matter to draw sharp lines between the civilian and military space-missile organizations; it was even less simple to draw sharp lines between the civilian and military space programs. The areas of overlap were very large and gray. The confusion inevitably resulting from the overlays of agencies and projects became greater as the international situation kept alive the question of whether space was primarily a civilian responsibility to be used for peaceful purposes or primarily a military responsibility to provide national defense. Under the circumstances there were endless opportunities for disagreements and rivalries that at any time might delay projects of vital interest to the United States. In the latter part of 1958 the situation was far from ideal, and it did not appreciably improve during the first six months of 1959.

VIII. EIGHT MONTHS OF ARPA SUPREMACY

When McElroy activated ARPA on 7 February 1958, he intended it to be either a "special task force" within the Department of Defense or possibly a "fourth service" to direct and control the research and development phase of the military space program. For at least a year it seemed that ARPA might indeed continue indefinitely to function as a fourth service, and during the first eight months of the period, February through September, it had a yet greater role, for it served as the civilian space agency as well. The President himself confirmed this temporary overall authority.[*]

The Sources of ARPA's Program

In the hectic days after Sputnik, civilian authorities had not only to determine high policy--questions of space-for-peace, of a single or dual space program, of space agency organization--but also the kind of projects to receive immediate emphasis. There were some who felt that neither Vanguard nor the Army-sponsored Jupiter-C proposal had any inherent value except as a "spectacular first," and Sputnik had robbed them of that. In the future the United States should forsake any project that smacked of "second best" and concentrate on another "spectacular first" as the only way to surpass Sputnik. Others argued that the United States

---

[*]In a memo to McElroy on 24 March the President approved the assignment of scientific and military space projects to ARPA, as the Secretary had set forth in a memo of 19 March. The President said: "I do so with the understanding that when and if a civilian space agency is created, these projects will be subject to review to determine which would be under the cognizance of the Department of Defense and which under the cognizance of the new agency."

needed psychologically to get into space at once, even though American satellites could not equal for the time being the success of the Russians. The latter argument prevailed.[1]

Vanguard and Explorer I

When reports of Sputnik first reached Redstone Arsenal, General Medaris and Dr. Von Braun, thought at once that their hitherto rejected Jupiter-C project might now be acceptable. They immediately briefed their guest, SOD Designate McElroy, and assured him the Army could place a satellite in orbit within 60 to 90 days. On 7 October, Wilbur M. Brucker, Secretary of the Army, recommended that the Secretary of Defense approve an Army program to launch a satellite within 120 days at a cost of $12.7 million.[2] This proposal was on McElroy's desk when he became Secretary of Defense on 9 October.

The new Secretary, assuming that a military satellite would be tolerated under the changing circumstances, asked Brucker to restudy the proposal and, at the same time, suggest ways of assisting Vanguard. The Army promptly discussed its project with Holaday, who was already reviewing plans to accelerate the Vanguard schedule in an indirect way. The Navy had planned to launch a Vanguard test vehicle (TV-3) late in 1957, but without intending to orbit it. After Sputnik the date was set for 6 December, the objective of the launching was changed to achieve orbit, and the test vehicle was advertised as a satellite. In view of these Vanguard plans, Holaday urged that the Jupiter-C not be launched until after 6 December. McElroy agreed, but on 8 November he authorized the Army to proceed, knowing that the Jupiter-C could not be launched before the Vanguard.[3]

On 3 November 1957 the Russians successfully orbited Sputnik II, the satellite weighing 1,120 pounds. One month later, 6 December, the Navy attempted to orbit the Vanguard test vehicle with its 3.25-pound payload. There was a mechanical failure in the propulsion system, and Vanguard burst into flames two seconds after launching. There was some criticism of the decision to turn a test vehicle launching into a satellite launching: "We pushed Vanguard too hard, and the project became a mess."[4]

Meanwhile, the Army went forward with its Jupiter-C project, now designated Explorer. It was being planned as a "scientific satellite," and objections no longer were there to using a military missile, the Redstone, as a booster. Launching occurred 31 January 1958, and a cylindrical satellite weighing 30.8 pounds with a perigee of 217 miles, an apogee of 1,093 miles, and an estimated life of 7 to 10 years was successfully orbited. The shot took place only 84 days after McElroy's authorization. The chief value of Explorer I lies in its irrefutable confirmation that the United States could have launched a satellite before the Soviet Union if the Army had received permission to make the try.

Explorer I was a great boon to Army prestige. In this connection it is interesting to note that the Army had the same advantage, in miniature, over the Navy and the Air Force that the Soviet Union had over the United States. The Army had an available missile with sufficient thrust to serve as booster in lifting a small satellite into orbit. Plans for Project Orbiter were based on that simple fact, and it was this same simple fact that led McElroy to give ABMA responsibility for the first successful American satellite. This did not imply that the Army had a carefully thought-out space program but rather that the Army met a national crisis in one field by using a weapon developed for another field as an emergency system of propulsion. Secretary Brucker said much this same thing in testimony before a congressional committee at a later date:[5]

> The Army developed its broad capabilities, which are now being used in space projects, as an inevitable result of its progressive work and outstanding success in the field of military missiles. The Army is gratified that as a result of this developed capability it can lend substantial support and assistance to the vital national space program.

Air Force Requests for DOD Approval

The Air Force was quite aware that McElroy's interest in Explorer could bode ill for USAF interests. On 29 October 1957, while OSD was still examining the Army's proposal, representatives of the Air Staff briefed the Secretary of Defense on the background and current status of the Advanced Reconnaissance Satellite (ARS) or WS-117L, pointing out that with a small increase in funds for fiscal year 1959 the satellite could be orbited in 1960. During the first two weeks of November the Air Force submitted suggestions to the Armed Forces Policy Council and to the Secretary of Defense that a Thor-boosted recoverable photographic satellite be launched in March 1959, that 12 Navaho boosters be utilized in various combinations to orbit satellites with payloads varying from 75 to 2,000 pounds, and that payloads of 28 to 270 pounds be sent to the moon within the next 8 to 12 months. On 12 November 1957, Secretary of the Air Force Douglas requested the Secretary of Defense to assign to the Air Force responsibility for all military satellites including, of course, WS-117L which was to be accelerated. There was no answer to the Air Force papers prior to the launching of Explorer I.[6]

Recommendations of the Advisory Group on Special Capabilities

On 6 September 1957, Holaday had written a memorandum to Dr. Stewart, chairman of the Advisory Group on Special Capabilities, evaluators and selectors of the IGY satellite:*[7]

> You and the members of your group constitute a unique body of experience in satellite systems and the actual problems of their development, having studied the possibilities before the announcement of the scientific satellite program, and having monitored its progress from the beginning. You also considered the larger, longer range possibilities such as WS-117L early in 1956 and made certain recommendations on it to the Department of Defense which for a number of reasons could not be implemented at that time.
>
> I feel that it may be timely to ask the group to help me by preparing for the time when it will ultimately be necessary to decide on a number of questions on military applications of satellite techniques. The feasibility and timing of such applications seem to depend mainly upon the capabilities of rocket systems, their availability, and, of course, upon the outcome of Project Vanguard, our first venture in this field.
>
> I should now therefore like to ask the Advisory Group on Special Capabilities to look again into the satellite plans and programs of the military departments and submit your conclusions on the technical capabilities based on the best available facts at this time . . . . As to timing, I shall be grateful if you could submit your main conclusions by March 1958.

The group held a meeting on 3 October, and the next day its leisurely approach was disrupted by Sputnik. Holaday requested that the group "expedite its study in every possible way," and on 11 October the group asked each of the three military services to submit recommendations as soon as practicable.[8]

The services replied in December with what may be called their first official programs. They agreed that the Russian success should be countered by a U.S. national program that integrated scientific and military elements "to avoid a dilution of effort," and they all looked toward manned space vehicles as the chief goal for the future. Otherwise each service thought along the lines of its own traditions. The Army and Navy,

---

*See above, p 72.

being surface-minded, wanted the space program designed to support land forces and fleets. The Air Force, on the contrary, thought of space as an extension of the operational area of airpower.

The Army's immediate interest covered reconnaissance, meteorology, mapping, geodesy, and navigation. Beyond this there should be deeper probes into the solar system. In a three-year program, the Army suggested 16 Jupiter launchings that would provide a 20-pound reconnaissance satellite by mid-1958, a 15-pound lunar shot by September 1958, a 120-pound lunar shot with photography by January 1959, and a 50- to 100-pound lunar impact sometime in 1959. Stretching out another dozen years to 1971, the Army spoke of manned carriers propelled by Titan-like boosters with sundry combinations of high-speed stages. The estimated cost was $14 billion. The Army was strong in its opposition to a single-service military program, though advocating a unified program to meet the legitimate needs of all three services. The Army also opposed recognition of space operations as an extension of strategic air activity.[9]

The Navy stressed small satellites, not exceeding 300 pounds, to meet immediate military requirements for communications, navigation, meteorology, and reconnaissance. They could be launched either by improved Vanguards or Thors in a schedule of 50 vehicles through 1961. Small, 10-pound satellites could also be advantageously launched from flying aircraft. The long-term program included manned vehicles of the X-15 type, five lunar shots, and eventually 1,500-pound satellites using Titan-Vanguard combinations for propulsion. The cost was considerably but vaguely more than $212 million.[10]

The Air Force suggested a short-term Thor-boosted 300-pound recoverable photographic satellite by 1959. A long-term program, based on Atlas and Titan missiles in connection with WS-117L, included missions for photography, ferret detection, infrared surveillance of aircraft and ICBM's, and eventually visual surveillance with television. With program acceleration, a WS-117L 2,000-pound satellite on a 300-mile orbit would be possible by 1959. The great advantage of WS-117L was that it had been under study since 1946 and under development since 1956. The weakness of the Air Force paper was its lack of cost estimates.[11]

The advisory group recognized valid military, scientific, and, eventually, commercial needs for satellites and believed that the objectives of the overall program would have to include manned spaceflight drawn from the X-15 experience. The group urged both an immediate short-range as well as a long-range program that would reach toward the genuinely spectacular. There were four major recommendations. First, plan for a strong program of large satellites and manned flight. Second, take immediate action to use the available potentialities of Vanguard and Jupiter-C[*] to launch very small satellites, as well as the Jupiter and Thor IRBM's to launch 300- to 400-pound satellites by 1959. This recommendation called for a Thor-117L interim program while anticipating the Atlas-117L. Third, WS-117L should be continued and given both military and nonmilitary application.[†] The fourth recommendation urged that the scientific parts of the

---

*Already Jupiter-C was being called Juno I, a Redstone booster with three clusters of Sergeants. Juno II was a Jupiter booster with three clusters of Sergeants, and Juno III was a Jupiter booster with three clusters of Vanguard Stage 3.

†In evaluating WS-117L the group noted the limited support given the project to date and emphasized the feasibility of using both the Thor- and Atlas-boosted combinations for such nonmilitary and military purposes as pure scientific exploration, communications, weather forecasting, etc., in addition to the planned reconnaissance and surveillance tasks.

national program should be carefully related and "mutually reinforced."[12]

The USAF Astronautical Program of 24 January 1958

Anticipating the Stewart Committee's report by several days, Holaday requested the Air Force--and presumably the Army and Navy too--on 7 January 1958 to suggest ways of expediting the space effort. Holaday speccifically stated that the purpose of the paper was to assist the Director of ARPA during the coming period of his indoctrination.[13] The Directorate of Research and Development prepared a summary statement on the Air Force astronautical development program, listing 5 systems and 21 subsystems to carry out 6 types of missions "essential to the maintenance of our national position and prestige." (See Table, p 132.) Two areas were mentioned as being of interest to both the military and scientific programs--space research and manned flight. Four other areas--reconnaissance, weapon delivery, data transmission, and countermeasures--were considered of military interest only. The program covered a period of 10 years, and the cost was estimated as an additional $61 million for fiscal year 1958 and $1.2 billion for fiscal year 1959. Assistant Secretary Horner forwarded the proposal on 25 January and requested Holaday to approve it and grant the required resources.[14]

There was an unfortunate misunderstanding within Air Force circles about the purpose of Holaday's request. DCS/Development and the Director of Research and Development thought the program should remain in and be carried out by the Air Force. Horner, too, seemed to have made the same assumption, else his request that Holaday approve the program and grant the requisite funds was scarcely comprehensible. Holaday, on the other hand, used the paper as he said he would--to assist Johnson during his

The Air Force Astronautical Program of 24 January 1958

Program	Subdivision	Mission
I. 609 Ballistic Test & Related Systems	1. BRATS 2. Aerial survey and target locating system	Space research Reconnaissance
II. 447 Manned Hypersonic Research System	3. X-15 4. Advanced hypersonic research aircraft	Space research Manned space flight
III. 464 Dyna Soar	5. Manned capsule test 6. Conceptual test 7. Boost glide tactical 8. Boost glide interceptor 9. Satellite interceptor 10. Global reconnaissance 11. Global bomber	Manned space flight Manned space flight Weapon delivery Countermeasures Countermeasures Reconnaissance Reconnaissance
IV. 117 Satellite Systems	12. ARS & photo capsule recoverable data 13. 24-hr reconnaissance system 14. Global surveillance 15. Manned strategic station 16. Strategic communications station	Reconnaissance Reconnaissance Reconnaissance; space research Weapon delivery; reconnaissance Data transmission
V. 499 Lunar system	17. Manned variable trajectory & test vehicle 18. Nuclear rocket test 19. Ion propulsion test 20. Lunar transport 21. Manned lunar base	Manned space flight; Space research Space research Space research Manned space flight; Space research Weapon delivery; reconnaissance

indoctrination. Holaday therefore made no reply to Horner, which was keenly disappointing to General Putt and his staff. There were some who felt that the program had been pigeonholed to die--to be "overtaken by events," as was said occasionally of later Air Force proposals.[15]

Last USAF Efforts to Save the Astronautical Program

By the end of January it was plain that McElroy would activate ARPA within a few days. It was also plain that ARPA would take over the military space research and development program unless McElroy could first be persuaded to reconsider the move. The Air Force then made three last attempts to limit ARPA's authority.

On 1 February 1958, Douglas harked back to his memorandum of 12 November--still unanswered--in which he had asked that the Air Force have responsibility for WS-117L satellites. Now, more than two months later, he again addressed the Secretary of Defense, requesting that the latter approve a draft paper containing the following paragraph and return it to the Air Force as a directive:[16]

> In connection with the proposed establishment of ARPA, of which you are aware, I desire that the foregoing project for a military reconnaissance satellite, as accelerated in the proposal submitted to me under date of November 12, 1957, be continued by the Air Force. However, no significant changes should be made in the program as so approved without the specific approval of ARPA. Pending the definitive establishment of ARPA the Director of Guided Missiles will have directional authority in respect to the program.

Again there was no reply, and on 7 February McElroy activated ARPA.

For several weeks Johnson was too busy setting his new house in order to exercise authority over the services. In this moment of respite, Douglas felt there was still the possibility of saving the integrity of the Air Force program. Once more, on 14 February, he approached McElroy and requested authorization for the Air Force to undertake five projects closely related to, but more detailed than, the Thor-WS-117L proposals of the previous November and December. These included ICBM nose-cone testing using a Thor-Vanguard combination; a Thor-Hustler television satellite, to be launched in September 1958, primarily for weather forecasting; a Thor-Vanguard satellite, with a first-flight date of July 1958, to carry out reentry experiments; a Thor-Hustler scientific satellite to be launched initially in October 1958; and a Thor-Vanguard launching for the purpose of hitting the moon.[17]

A week passed with no acknowledgment from the Secretary of Defense, and on 21 February the Air Force made another try to save WS-117L for itself. This time Horner requested the Secretary of Defense to designate the Air Force as executive agent for WS-117L since its development plan was already being readied. Indeed, a contract with Lockheed was being supported from Air Force resources during fiscal year 1958. Provision for the contract had been included in the original fiscal year 1959 budget, but in the course of formulation the funds had been deleted in favor of ARPA. Horner hoped these funds would be returned to the Air Force, with authority to proceed.[18]

When McElroy replied on 24 February, he ignored Douglas' requests of 1 and 14 February. However, the Secretary of Defense approved the acceleration of WS-117L, but under the direction of ARPA. Also, he requested that a fund status summary of Air Force space projects be submitted to ARPA.[19] The Air Force knew that development responsibility over USAF space projects had passed to ARPA. Of course Headquarters USAF prepared the financial statement and submitted it next day. (See summary, p 135.)

Air Force Fund Status Summary
February 1958
(In Millions)

Projects Under Way	Programmed FY 1958	Funds FY 1959	Add FY 1958	Rqmts FY 1959
BRATS study (not yet approved by OSD)	3.60	.50	20.00	177.50
X-15 study	29.30	18.90	7.00	155.20
Dyna Soar	3.66	6.00	8.90	177.50
ARS (FY 1959 funds in ARPA budget)	48.05	(96.00)[a]	2.00	245.00
Lunar base studies	.70	.40	3.60	80.60
Technical development (including 1-million-lb-thrust rocket, human factors engineering, electronic techniques, & atmospheric physics)	90.00	110.90	14.50	214.10
Basic research (propulsion, materials, geophysics, etc.)	28.10	30.10		44.20
Test & instrumentation	8.20	10.10		27.30
Center operations			5.00	30.00
Projects submitted OSD for approval Television satellites Recoverable satellites Scientific satellites Moon impact				
TOTAL	211.61	176.90	61.00	1,151.40

[a] Requested in the ARPA 1959 budget and not included in totals.

(Memo, B/Gen H.A. Boushey to C/S USAF, 28 Feb 58, subj: Status of USAF Astronautics Program, w/incl, Data Sheets.)

The Rule and Program of ARPA in 1958

McElroy had ample opportunity in his first months in office to become familiar with what the services had to offer and what they desired for the still nebulous space program. Before ARPA was activated in February 1958 the Secretary experienced the disappointment of the Vanguard attempt on 6 December and another Vanguard failure on 5 February. On the other hand, he was doubtless encouraged by the success of Explorer I on 31 January and by plans for the continuation of that project. He had learned from briefings, memorandums, conferences, and reports the potential capabilities of the Army, Navy, and Air Force in space activities, and he had also the recommendations of the Stewart Committee to guide him in selecting projects for assignment to ARPA.

ARPA's Operating Procedures

Johnson's approach to the services was not altogether a happy choice. Although ARPA was established ostensibly to direct the research and development phase of military space projects, the activating DOD Directive No. 5105.15, of 7 February 1958, was couched in general terms: "The agency shall be responsible for the direction or performance of such advanced projects in the field of research and development as the Secretary of Defense shall, from time to time, designate by individual category." More specifically, the agency was authorized to direct the assigned projects, whatever they might be, by contractual arrangements with both government and nongovernment agencies. Also, ARPA was authorized to acquire or construct facilities as necessary. The Secretary of Defense thus remained in a position to control the growth and responsibilities of ARPA by either limiting the agency's responsibility to individually assigned projects or granting an overall authorization for wide areas. The services of course could do nothing but follow a "wait and see" policy while ARPA's true significance slowly unfolded.

There was one thing, however, of which both ARPA and the services could be sure--in the specific areas of its eventual assignments, whatever they might be, ARPA would possess an authority superior to that of the Army, Navy, or Air Force. For more than a month, Johnson did very little to show what operating procedures he would employ. Then on 27 March 1958 he sent nearly identical memorandums of basic policy to each of the service secretaries. Though authorized to do so, he would not in the near future construct or acquire facilities, but he asserted his right to take over service laboratories whenever he should deem it advisable. After ARPA received project assignments from the Secretary of Defense, he would reassign them among the services or perhaps outside the services--whichever might be conducive to greater efficiency. In pursuit of ARPA objectives, Johnson stated that he was free to deal directly with field agencies, completely bypassing service and command headquarters. He listed the Army Ballistic Missile Agency, the Air Force Ballistic Missiles Division, other centers of the Air Research and Development Command, and the Naval Ordnance Test Station (NOTS) at Inyokern, Calif., to which he would issue directives from time to time "for technical and administrative services."[20]

The services did not relish Johnson's decision to act independently. There was no question of his authority, but there was a question of the wisdom of his decisions. To have a service project assigned to ARPA and then have it splintered into components for reassignment among service or outside agencies might lead to increased efficiency in the development of some parts, but it seemed unlikely that the project as a whole would benefit from dismemberment. Moreover, his policy of suspending established methods of communication could lead to confusion, and it seemed to be at least a partial transfer of indispensable field units from the control of service headquarters to a fourth service. It was rather widely assumed that this disregard for normal channels of communication came from Johnson's IDA advisers who were "inexperienced with military methods of procedure."

There is always another side of the coin. Johnson had great authority, but with it came corresponding difficulties. Though he could indeed act independently as if chief of a fourth and superior service, his position within the Department of Defense made it necessary for him to act as arbitrator in service differences over space. It is also right to mention that Johnson did not get the idea of out-of-channel communications from his IDA advisers but from Secretary of the Army Brucker, who suggested it as a "time-saving" device.[21]

The out-of-channel communications did not work well. It nevertheless required most of 1958 for ARPA to concede and make amends. In the end, ARPA decided first to deal directly with ARDC rather than AFBMD and, somewhat later, to recognize the rights of Headquarters USAF.[22]

The Assignment of Projects to ARPA

Two months elapsed before any projects were actually assigned to ARPA, but Johnson did not wait that long to assert his authority in the area of space research and development.

McElroy's memorandum to the Secretary of the Air Force on 24 February 1958 stated definitely that WS-117L would be placed under ARPA. Four days later, Johnson gave evidence of his right to speak for the Secretary of Defense in matters pertaining to space. Expressing his own interest in USAF's long-term claims to manned space flight and WS-117L, Johnson said the Air Force should concentrate on these two fields even to the detriment of lower-priority projects. He wanted WS-117L accelerated but rejected the interim Thor-boosted version in favor of the Atlas version and requested a clarification of the whole Air Force program.[23]

Accordingly, Air Staff representatives briefed Johnson on 19 March 1958. The briefing covered unmanned systems, Dyna Soar, lunar base, and manned satellites as a substitute for manned hypersonic research system. Explanations of WS-117L were limited to the Atlas version, and MIS was still kept to the employment of a manned capsule in preference to concentration on Dyna Soar. Only the capsule method was considered capable of putting a man in space ahead of the Russians.[24]

That same day, Johnson asked SOD-Presidential approval of three space projects selected by ARPA. Project No. 1, to be assigned to ABMA, called for launchings--in August, November, and December 1958 and January 1959-- of a high-visibility "propaganda" satellite, an escape guidance experiment, an IGY satellite, and a cloud-cover experiment. Project No. 2, to be assigned to AFBMD, consisted of three lunar probes using a Thor booster, part of Vanguard as second stage, and a solid rocket as third stage. Project No. 3, to be assigned to NOTS, was the development and operation of a mechanical ground-scanning system for the lunar probes.[25]

McElroy sanctioned the projects within hours and forwarded Johnson's request to the White House. Five days later, the President signified his approval but carefully made the point that only for the time being ARPA was acting as the national space agency. Upon the activation of a civilian agency, he warned, there would be a reevaluation and redistribution of projects. On 27 March 1958, Johnson was thus in a position to issue ARPA Orders (AO's) Nos. 1, 2, and 3 to ABMA, AFBMD, and NOTS to undertake the development of the space vehicles, the lunar probes, and the scanning system.[26]

The buildup of the national space program began with AO's 1, 2, and 3. Other projects passed to ARPA in quick succession. On 4 April, Argus, the high-altitude atomic effects tests scheduled for the South Atlantic area in the near future, became an ARPA responsibility though not technically part of the space program.[27] On 1 May another transfer was made by an OSD directive that stated:[28]

> . . . all satellites and other outer space vehicle programs to be conducted by the Department of Defense, including the VANGUARD series, are hereby reassigned from the Director, Guided Missiles, to the Director, Advanced Research Projects Agency. The VANGUARD reassignment specifically includes responsibility for preparation of the monthly reports to the President on the progress in the International Geophysical Year Satellite programs.
>
> The Director, Guided Missiles, will continue to be responsible for support of the above programs by necessary rocketry, launching and other range facilities, and the like.

By the time that Vanguard became an ARPA project there was obvious need of a systematic way to record the transfers. On 17 May 1958 the Department of Defense issued Directive No. 3200.5, which repeated the February definition of ARPA's authority and also served as the basic paper to which all future transfers to ARPA would be recorded as inclosures. AO's 1, 2, and 3 remained separate, but the Argus and Vanguard transfers automatically became Inclosures 1 and 2. Between then and October 1958 numerous other assignments were made, including WS-117L on 30 June. The assignments covered the whole spectrum of "space-related" projects from propellants to engines, electronic vehicles, tracking, defense against ballistic missiles, and satellites and space probes.*

By the first of July it was possible to place ARPA's projects into three broad areas: ballistic missile defense, chemical propellant research, and military space, the last for the time being including some projects destined for the civilian space agency. The frontiers of ARPA had been drawn.

ARPA's Assignment of Projects

The distribution of projects among the services began with AO's 1, 2, and 3. The system of formalizing the assignments was satisfactory and underwent no change. Between 19 March and 1 October 1958, ARPA issued 22 AO's, which in turn were subject to numerous amendments from time to time.

It was soon evident that Johnson was following through on his announced policy of assigning, and even splintering, projects among the services and other agencies as he saw fit. The Air Force, in a last attempt to preserve the integrity of its program, decided upon a new tactic. Rather than appeal uselessly to the Secretary of Defense, an appeal should be made

---

*Projects transferred to ARPA, and the dates:
1. Argus (nuclear explosions in exosphere over South Atlantic) 4 Apr 58
2. All DOD approved satellite and outer space programs (with Vanguard) 1 May 58
3. High-performance solid propellants 7 Jun 58
4. Minitrack doppler fence 20 Jun 58
5. USA and USAF ballistic missile defense projects except Nike Zeus and BMEWS 20 Jun 58
6. Studies of effects of space weapons employment on military electronic systems 20 Jun 58
7. Nuclear-bomb-propelled space vehicle 20 Jun 58
8. Superthrust rockets 20 Jun 58
9. WS-117L 30 Jun 58

to Johnson to assign the Air Force a revised USAF integrated program instead of its dismembered parts.[29] In April, Headquarters had a plan consisting of four projects--an accelerated WS-117L Advanced Reconnaissance Satellite, to be operational by March 1960; a man-in-space capsule; a manned lunar base for intelligence observations of Earth and outer space; and the continued development of the 300,000- to 400,000-pound rocket engine begun in 1954, the 1,000,000-pound single-chamber engine begun in 1957, and the AEC-USAF-sponsored nuclear-bomb-propelled vehicle. Maj. Gen. Jacob E. Smart, Assistant Vice Chief of Staff, forwarded the proposed plan to Horner for transmission to Johnson.[30] The memorandum remained unsigned on Horner's desk until mid-June. By that time it had been "overtaken by events" and was returned to General Smart.

In the meantime, ARPA's breakdown of programs and projects, and their reassignment, continued. The Air Force received back from ARPA, on a contractual basis with ARDC and AFBMD, studies of satellite defense, effects of space weapons on electronics, and the feasibility of nuclear-bomb-propelled space vehicles. In addition, the Air Force also received assignments for research in high-energy fuels, the development of WS-117L and Project Score. The latter was a propaganda stunt to send a complete Atlas vehicle into orbit, equipped to broadcast a recorded Christmas message to "the world" from the President.

The Air Force was pleased to have these assignments even as contracts between ARPA and ARDC or ARPA and AFBMD. Nevertheless, the Air Force was seriously disturbed by ARPA's persistent splintering of projects into components, as for instance separating the three lunar shots from the development of a mechanical system to track them. The Air Force felt a keen sense of loss in giving up to the Army the projects for the cloud-cover satellite and the new 1.5 million-pound clustered engines. By midsummer the identity of the well-thought-out Air Force space program had been lost, the projects either assigned back to ARDC or AFBMD under ARPA management or scattered among other agencies. Indeed when the Directorate of Advanced Technology came into being on 15 July 1958, the director, General Boushey, had little to direct other than seven studies in a space study program devised since January:[31]

Number		Date of Origin	Objective
1.	SR 178	12 Feb 58	Global surveillance system to determine design of manned reconnaissance satellite system.
2.	SR 181	10 Jul 58	Strategic orbital system to determine concept for military operations in Earth orbital space.
3.	SR 182	25 Jul 58	Strategic interplanetary system to determine military usage research for vehicle and test.
4.	SR 183	4 Apr 58	Lunar observatory as approach to manned observatory on the moon.
5.	SR 184	24 Apr 58	24-hour reconnaissance satellite for continuous surveillance of preselected areas on Earth.
6.	SR 187	1 May 58	Satellite interceptor system to combat hostile satellites with early detection and elimination.
7.	SR 192	29 Aug 58	Strategic lunar system to determine feasibility of using the moon for military purposes.

There was another matter resulting from the distribution of projects that caused concern in Headquarters USAF by the late spring of 1958. The successful orbiting of Explorer I on 31 January--followed on 17 March by the successful orbiting of Vanguard I--placed the Air Force in an unfavorable position. The two satellites were most gratifying as accomplishments, but they left the Air Force as the only service that had not demonstrated an ability to launch a satellite despite the claim to preeminence in space. Headquarters feared the situation would become "even more embarrassing" in the next few months. The only promise of an early success, after ARPA began distribution, was in the three lunar shots specified in AO No. 2.

In April and May, Headquarters USAF thought of the lunar shots optimistically. There seemed a likely chance of success for either the first or second try. In that event, the third shot, as an unnecessary duplication, could probably be placed under WS-117L as a biological experiment.[32] Unfortunately, that is not the way the lunar probes turned out. The first shot, on 17 August 1958, reached an altitude of only 40,000 to 70,000 feet. Before the second and third shots could be fired, the project passed from ARPA to NASA, which, like ARPA, operated in this instance through AFBMD. The second shot, on 11 October, reached an altitude of 70,700 miles, and the third shot, on 8 November, went only to 963 miles.

The year 1958, whether during the period of ARPA's supremacy or after the division of the program with NASA, was not turning out well for the Air Force. Not until 18 December did Air Force competence prove itself in space endeavors, for it was then that the Atlas missile of Project Score went into orbit.

IX. NASA'S FIRST PROGRAM, OCTOBER 1958 TO JULY 1959

During the first half of 1958 the services lost managerial control of their space projects--but only to ARPA. At the same time it was evident that the activation of NASA, set for the early autumn, would bring a day of reckoning, a day for the division of the national program between ARPA and the new civilian agency.

The question that remained undecided during crucial weeks was where to draw the dividing line between the military and civil programs. There were large areas of overlapping interests, projects which were of importance both for strategic and for scientific reasons. It would be a simple matter, of course, to make arbitrary decisions, to say which projects were to be kept within the military program and which were to be transferred to civilian control. But arbitrary distinctions between military and civilian programs might not be wise from the viewpoint of national interest. The nature of World War II and the international situation that existed after the war blurred the lines between civilian and military activities. Astronautics, whether civilian or military, whether aimed at preparedness or peaceful purposes, would make important contributions to human welfare, to the political prestige of the United States, and to the defensive and offensive strength of the nation.[1]

The Space Act of 1958 provided both for military and civilian space programs and for cooperation and coordination between the space agencies. The aim was to avoid undue duplication. The danger was in assignment of borderline projects to civilian management. Despite the best intentions to cooperate, the civilian agency would be motivated by scientific objectives devoid of the urgency required by defense. Under the circumstances it seemed better to keep the border projects under military control or, failing this, to tolerate some duplication rather than hold back the military use of vehicles for which there was no pressing need among civilians.

ARPA's Claim to Border Projects

From the President's message to Congress on 2 April 1958 it was clear that NACA would become the nucleus of NASA, and there would be a wide overlap of civilian and military interests. With ARPA already serving as the national space agency, it was expedient for Johnson, Director of ARPA, and Dryden, Director of NACA, to establish a "jurisdictional committee" to determine as far as possible the ARPA and NASA areas of operations. Negotiations were in progress before the end of April. Johnson made a strong effort to keep military losses to the minimum. As pro tempore head of the national space program, he organized ARPA's existing projects into four categories. Category I, Defense vs ICBM's, covered the entire field except Nike-Zeus and BMEWS. By their very nature, there was no chance of Category I projects being transferred to NASA. Category II, Military Reconnaissance Satellites, was little more than the Air Force WS-117L program, and that too was certain to remain under DOD control. Category III, Military Developments for and Applications of Space Technology, was "a collection of smaller items," which became the real bone of contention between ARPA and NACA (NASA). As Dr. York said before a congressional committee on 23 April, "On our first go-round, and for our own part, we /in ARPA/ list all these /Category III projects/ as being military developments."* Category IV, indisputably destined for NASA, included the four satellite experiments of AO No. 1 being designed by the Army; the three lunar probes of AO No. 2 entrusted to AFBMD and their payloads being developed by NOTS; Project Vanguard; and the Explorer series for cosmic ray, solar, and astronomical measures as well as meteorological and biological research.[2]

By the first week in May, ARPA and NACA agreed that the initial program for NASA would contain three principal areas of interest: use of unmanned space vehicles instrumented to collect scientific information; development of science, technology, and equipment required for manned space flight; and research and development of components and techniques needed to increase the national capability in space technology.[3]

Throughout the remainder of the spring and well into the summer of 1958, ARPA continued to hope and work for a strong military program. The basic philosophy was that the United States could not permit, either from a national or military standpoint, a foreign power to control space. This condition overruled the argument that the military should attempt no space exploration until it was possible to determine specific military usefulness. As ARPA spokesmen pointed out, "A strong military research and development program that will lead to manned and unmanned space orbiting weapon systems and space flight vehicles to permit military

---

*Category III included man-in-space; Operation Argus (the high-altitude tests of atomic detonation effects held in the South Atlantic in the autumn of 1958); satellite tracking and monitoring systems (which had come to the fore in early spring discussions of celestial traffic control); satellite communications relay, meteorological reporting, and navigational aid systems; bomb-powered rockets; and solid propellants. These were projects of mutual scientific and military interest, and the services hoped to keep them in DOD.

operation in space can be the key to future national survival."

ARPA believed that military space missions would fall into four types: defensive missions to defend the United States from ICBM's, IRBM's, and satellite weapon carriers; offensive missions for purposes of deterrence or strategic weapon delivery; information missions for surveillance, communication, weather observation, and space traffic control; and space bases for logistic purposes.[4]

The President's Division of Space Projects

Actually the military were fighting a lost cause. Under the Space Act of 1958 the President had authority to determine which agency should be responsible for specific projects. His policy of space-for-peace made him reluctant to grant the military any space activity that could be considered of scientific interest, and when he signed the Space Act on 29 July, he made it clear that borderline projects would go to NASA.[5]

Two months later, on 1 October, when Dr. Glennan activated the civilian agency, the President confirmed this decision in Executive Order 10783. He thereby transferred to NASA responsibility for:

    (a) The United States scientific satellite project (Project VANGUARD)
    (b) Specific projects of the Advanced Research Projects Agency and of the Department of the Air Force which relate to space activities (including lunar probes, scientific satellites, and superthrust boosters) within the scope of the functions devolving upon the National Aeronautics and Space Administration under the provisions of the National Aeronautics and Space Act of 1958, and which shall be more particularly described in one or more supplementary Executive orders hereafter issued.

With the exception of Project Vanguard, however, the specific projects entrusted to NASA were not defined since the "one or more supplementary Executive orders" were never issued. The omission enabled NASA to claim practically any ARPA or USAF project remotely connected with astronautics provided the President did not disapprove. In October 1958, ARPA lost control of Category IV, as had been expected, and also a large part of Category III--which may also have been expected but was certainly contrary to military hopes. Still worse, it appeared probable that the remainder of Category III might also pass to NASA in the near future.*6

Along with its sweeping responsibility for space projects, NASA acquired extensive but scattered facilities. With the absorption of NACA, it acquired the three research centers at Langley AFB, Va.; Cleveland, Ohio; and Moffett Field, Calif. In addition there was Wallops Island, Va., and several other field offices. Within a matter of weeks the President also transferred to NASA most of the Army's interest in the Jet Propulsion Laboratory under the California Institute of Technology. At the same time, that is within the first two or three months of its existence, NASA made at least two vain attempts to get part of the Army Ballistic Missile Agency at Huntsville, but the President's reputed reply of "not at this time" only postponed the answer ARPA feared would be affirmative sooner or later.7

NASA's First Nine Months

NASA's responsibility was to organize a space program considerably broader than that required by ARPA or the three services. In undertaking the work, Dr. Glennan, himself an outstanding scientist, had the assistance

---

*Actually, little of the original ARPA "space" program remained in the agency with the exception of WS-117L. But new projects were being discussed, and some were soon approved and placed under ARPA's authority, so that by the end of October the agency could boast an 11-project program.

of Dr. Dryden and the very competent former NACA staff. They were well versed in current space technology, and many of them had served on a special NACA committee that anticipated NASA's early needs by preparing a report, submitted to Glennan on 28 October 1958, to guide his first efforts. Entitled "Recommendations to NASA for a National Civil Space Program," the report emphasized the role that the physical and life sciences would play in the exploitation of space and pointed out the importance of manned space flight. It called for close coordination with "civil and military agencies" in related work, but, since NASA's program would be largely one of scientific interest, there was nowhere in the report that sense of urgency felt within the armed forces.[8]

Before the end of 1958, Glennan approved a NASA program for fiscal years 1959-60. (See program, p 151.) Among other things, it called for numerous space science and advanced technology projects, the latter with emphasis on the development of boosters and vehicles. In operations, the program would mean launching many sounding rockets, 35 satellites, 7 lunar probes, and 3 interplanetary probes before July 1960. It was an ambitious undertaking.

The space science program was very broad. It covered seven great areas of research: the atmosphere of the earth, moon, planets, and sun; the ionosphere; energetic particles; the electric and magnetic fields; the gravitational fields; astronomy; and biosciences.

Of course a program that reached so far into scientific investigation was necessarily one of unceasing flux. Details of projects, and entire projects, varied from month to month, almost from day to day. By the spring of 1959 the program in general content was about what it had

NASA's Program for FY's 1959-60

1. Supporting activities

    a. JPL funding-nondirected programs
    b. University contracts and grants--long-term nonspecific research

2. Space science program

    a. Sounding rockets, numerous launchings
    b. 35 satellites
    c. 7 lunar probes
    d. 3 interplanetary probes

3. Application program

    a. Meteorological satellite in connection with other agencies
    b. Communication satellite, to be at first a large balloon or reflective type
    c. 24-hour satellite to be developed with ARPA
    d. Navigational satellite but details very indefinite

4. Advanced technology program

    a. Vehicle technology
    b. Boosters--recoverable
    c. Propulsion
    d. Man-in-space (MIS)
    e. Human factors
    f. Scout

5. Ground support

    a. Tracking
    b. Launchings
    c. Guidance.
    d. Structures.

been in October, but the projects were more daring. NASA witnesses in congressional hearings spoke of lunar impacts, lunar orbitings, and soft lunar landings as well as manned space flight, all by the end of 1961. In addition, there would be deeper and deeper probes into space and the development of meteorological, communication, and astronomical observatory satellites. Experience was soon to prove, however, that optimism outran technology.[9]

If the space science program was to earn the aura of reality, it had to be based upon the possession, or at least the prospects, of adequate boosters and vehicles. Invested with authority by the Space Act of 1958 to call for cooperation from any Federal agency, Glennan cast aside completely the old taboo against the use of military rockets for the peaceful exploitation of space. He informed the President that military rockets would be used when needed and called upon DOD to supply him with requisite information, services, equipment, facilities, and personnel.[10] Since neither the President nor the Secretary of Defense offered objection to Glennan's action, it was evident that military rockets could then be used for scientific projects without travail of conscience.

The rockets immediately available to NASA were, however, of limited usefulness. Vanguard seemed still unreliable and, along with the more reliable Jupiter-C, offered small payload capacity. The combinations of Jupiter and Jupiter-C as well as Thor-Able were equally hampered by technical shortcomings that prevented high-altitude orbiting. At the end of the year NASA turned to Thor-Hustler (later redesignated Agena), a combination of Thor and Bell Aircraft's Hustler engine. It was the most powerful and most welcome member of NASA's first group of vehicles, but it could not meet all of the requirements of an expanding program.[11]

In January 1959, NASA listed three vehicles in a near-future second group and seven basic engines to be developed for the long-term program. The second group included a modification of Thor-Able with guidance to become Thor-Delta and two Atlas combinations, one of which, Atlas-Hustler, would lift a 3,000-pound payload. The seven basic engines included two modifications--one a Vanguard and the other a 6,000-pound storable-fuel JPL engine--and five new ones. The latter ranged in thrusts from the 15,000-pound Pratt & Whitney to the 1.5 million-pound single-chamber Rocketdyne.

NASA devised various combinations of the seven engines to provide four basic vehicles--Vega, Centaur, Saturn, and Nova. Saturn, being developed by ARPA-ABMA, would have a cluster of eight 188,000-pound engines with a total thrust of 1.5 million pounds, and Nova would have a cluster of four 1.5 million-pound single engines with a total thrust of 6 million pounds.[12]

Both the plans for the second group and for the seven basic engines were very impressive and constituted "a great leap forward"--at least on paper--provided, of course, that delays could be avoided on the way. NASA believed that between the fall of 1959 and the winter of 1961 it would be possible to increase orbital payloads from 300 pounds at altitudes of 300 miles to 800 pounds at 22,000 miles. The payload for the long-range prospects was estimated to be 8,000 pounds at 300 miles.[13]

As part of the overall propulsion program, NASA included more futuristic projects. A number of nonchemical systems were based on nuclear and electric engines; the latter included both ion and plasma rockets.

But no matter how promising they might be, the application of results would be part of very long-term objectives.[14]

In April 1959, Dr. Glennan stated that NASA scientists were "hard at work on problems connected with all our major military missiles--problems concerned with warhead stability, stage separation, and high energy fuels, to name a few. It is no exaggeration to say that just about every U.S. aircraft and missile had benefitted importantly from NASA research."[15] Certainly the NASA program did help the military, but not as an unmixed blessing. NASA adopted a program with fine scientific objectives, but the instruments with which the program was to be put into effect were largely military projects transferred to NASA authority. NASA's space vehicles, whether already available or under development, were originated by the military, sometimes in cooperation with NACA. NASA's research work in fuels, and in nonchemical engines was a continuation of research begun by the military. NASA's man-in-space or Mercury project was originally an Air Force dream. Once these projects passed from military to civilian control, even though their development continued and under efficient management, the emphasis shifted. They had one significance for the military, another for NASA, and there was a tendency to slow down in just those aspects of the program in which defense was most interested.

The exploitation of space is costly. If the United States could not afford to support two programs so that neither would interfere with the other, there were some who felt that national security demanded the survival of military projects, even if that meant elimination of civilian participation.

NASA-USAF Relations

The creation of ARPA and NASA affected the Air Force unfavorably, since its future might well depend upon a space role. To have the space program taken over by ARPA was a serious blow, and to have the program again divided with NASA was yet more disturbing. Entirely outside the Department of Defense, the leaders of the civilian agency thought neither in terms nor interests of the military but pursued space flight and space exploration as ends in themselves. Yet national defense was at stake.[16]

Favored by the President as an expression of space-for-peace, NASA could often impose its will upon the Department of Defense and the military services. Furthermore, the Bureau of the Budget was the voice of the President in matters of government finance, and in some ways became the final arbiter in matters of space. It distributed its benedictions among space projects and between the two space programs in accordance with the White House philosophy of economy and preference for the civilian agency.[17]

The Air Force hoped for cordial relations with NASA as a matter both of national and service interest, perpetuating if possible the excellent cooperation that had always existed between itself and NACA. The space agency seemed destined to play the major role in the American program for years to come, and the future of the Air Force lay in space. Cooperation could be beneficial to both. The Air Force could assist NASA with supporting facilities and experienced personnel, and NASA could assist the Air Force in projects of mutual interest.

There were of course occasions of misunderstanding, but the Air Force kept its goal of cooperation. This policy was brought out clearly in connection with the long-delayed WS-609 Ballistic Missile Test System

(BMTS). In the summer of 1958 the Directorate of Research and Development suggested a joint USAF-NACA effort for the project, and the subsequent negotiations between the two agencies culminated in a USAF-NASA Memorandum of Understanding on 31 October. It provided for USAF-NASA cooperation in the development of a solid-rocket test vehicle that the Air Force soon redesignated as the Hypersonic Environment Test System (HETS) and NASA called Scout.[18]

The Air Force hoped that the cooperation shown by NASA in connection with HETS-Scout would prevail generally, but unfortunately there were other signs of strain. In addition to NASA's no-urgency attitude and non-military security precautions, there were two major sources of irritation--NASA's interruptive demands on USAF facilities and resources used for ballistic missiles and NASA's tendency to assume proprietary rights in the lunar system of the USAF space study program.[19]

NASA's first contact with the Air Force came shortly after the agency's activation and was essentially in matters of logistics. Although the space agency inherited the facilities of NACA, these were inadequate to the vastly expanded requirements of the space projects. NASA thereupon proceeded to contract missile industry and civilian research centers and to take over facilities owned by or contracted to other government agencies.[20]

By January 1959, NASA had acquired the Jet Propulsion Laboratory, had tried to take over ABMA, and was writing contracts with Rocketdyne and the Space Technology Laboratories. Simultaneously, NASA received space projects previously assigned by ARPA to AFBMD. The transfer automatically broadened NASA's grip, with some disruption of the USAF ballistic missile program. The question for the Air Force, and for ARPA too, was how to share the missile-boosters with NASA and how to determine the NASA-service demands on missile industry and test facilities without serious impairment of the military weapon system program. As early as January 1959 the NASA-ARPA requirements for Atlas boosters generated major problems in scheduling. Also the launching of NASA space vehicles threatened to overtax the Air Force facilities at Patrick and Vandenberg.[21]

Thus at the beginning of the new year, NASA's enthusiasm, plus its failure to appreciate military requirements, imposed strains on relations with the Air Force. To ease the situation, Headquarters USAF at the behest of AFBMD sought a NASA-USAF agreement to apportion the demands being made on USAF resources for ballistic missile research, development, production, and testing facilities. In turn, NASA professed fear that such an agreement would interfere with the civilian space program and preferred to negotiate at the level of the Secretary of Defense. At the end of fiscal year 1959 there was still no NASA-USAF agreement on these vital questions, and the military program felt the disadvantage of having a lower priority than that of the civilian program.[22]

In the weeks following ARPA's activation, when service projects were being transferred to that agency by OSD decree, the Air Force began a space study program. Its inspiration was a desire by Air Force leaders to avoid in the age of space exploration the blind spot that had led to the lapse of the ICBM program between 1947 and 1954. The need was to look as far as possible into the future of space exploration and keep an integrated concept of possible operations ahead of current requirements. The means chosen to effect this aim was a relatively small program, already sponsored by ARDC,* with industry.[23]

Immediately after Sputnik, Headquarters USAF authorized ARDC to undertake a study of military space applications and of the research and development to attain and support the requisite systems. The conclusion was that there were three basic areas of permanent importance--Earth satellites, lunar control, and interplanetary exploration. In the next few months, February-August 1958, the ARDC-industry effort produced seven series of study requirements (SR's), six of which were organized as substudies under three strategic systems:[24]

Strategic Orbital System	Strategic Lunar System	Strategic Interplanetary System
SR 181, Strategic, 10 Jul 58	SR 192, Strategic Lunar System, 29 Aug 58	SR 182, Strategic Interplanetary, 25 Jul 58
SR 178, Global Surveillance, 12 Feb 58	SR 183, Lunar Observatory, 4 Apr 58	
SR 187, Satellite Interceptor System, 1 May 58		

The seventh series, the 24-hour reconnaissance satellite studies, SR 184, 1958, was regarded as a possible support system--along with the photographic satellite of WS-117L, the meteorological satellite, man-in-space, and Dyna

---

*The program was conducted on both a voluntary and funded basis. ARDC from time to time released to industry general descriptions of an area of probable future operational significance. Industry in turn undertook studies to determine the kind of weapons likely to be required, considering technical feasibility, operational concept, facilities, manpower, training, methods of development, production schedules, and overall cost estimates. The studies were then evaluated by ARDC, the School of Aviation Medicine, would-be interested commands, Rand, Headquarters USAF, and NACA. It was a complicated process, but the end results had been useful.

Soar--for the strategic orbital system.

The idea of a lunar base, which was carefully evaluated in SR 192 and SR 183, had already aroused strong interest in some USAF circles, and a lunar base had been listed as one of the five main divisions of the Air Force proposed astronautical program submitted to the Director of Guided Missiles on 25 January 1958. It was defended by General Boushey, Deputy Director of Research and Development, in congressional hearings during April, when he claimed that the moon could be used as a launching site for deeper penetration of space, as a supply base for earth satellites, as an astronomical and meteorological observatory, and as a means of worldwide surveillance that could be a deterrence to aggression. From time to time other spokesman said essentially the same thing and warned that it might well become a matter of urgency to claim the moon by landing there ahead of the Russians.[25]

The suggestion that a lunar base was militarily significant was challenged, but its scientific value could never once be denied. Naturally the strategic lunar system (SR 192) and lunar observatory studies (SR 183) excited interest among the NASA representatives at a NASA-USAF conference on 13 November 1958. NASA asked that it be kept apprised of the progress made by the whole space study program, particularly in the fields of overlapping interests. NASA wanted especially to know of the strategic lunar system status and, in return for this Air Force information, offered full reciprocation.[26]

In succeeding months, the strategic lunar system began to seem feasible with a logical extension of current techniques. In the spring of 1959 there was speculation that a manned lunar landing and return might be possible by 1967, and a permanent lunar base by 1969. The estimated cost was fixed at $8 billion and an annual operating cost after establishment of the base of approximately $600 million.[27]

In accordance with the agreement of November 1958, NASA was kept informed of progress but seemed less and less inclined to reciprocate. Gradually a background of unhappy incidents in NASA-USAF relations built up. In March 1959, ARDC invited NASA to participate in contractor midpoint briefings on SR 183. The response was markedly unenthusiastic, and only one NASA representative attended. Early in April, NASA created a Lunar Exploration Group. The Army and Navy had representatives on the group but not the Air Force. A short time afterwards, on 17 April, to the surprise of the Air Force, NASA announced plans for long-range scientific exploration of the moon.[28] It was at this same time that NASA representatives were speaking confidently of lunar orbitings and landings in their statements before congressional committees.

A few days later, at a scheduled Headquarters ARDC briefing by two contractors working on the strategic lunar system studies, NASA representatives "injected" the remark that the lunar area was "exclusively NASA property." This far-from-cooperative attitude by NASA in the lunar field became more noticeable as weeks passed, and it came to cover much wider areas. Although during the summer of 1959 NASA agreed to participate with ARDC in briefing the Department of State on space activities and programs, NASA soon reneged. The stated explanation was that NASA must avoid the impression of compromising its devotion to space-for-peace by seeming to associate its program with the military.[29] There were suggestions, however, that the Air Force's insistence upon urgency for the overall program was very irritating to NASA officials who, adjusting plans to budgets in the 1958-59 period, could not conceive of a lunar base except as a 20-year program.[30]

Whatever the explanation, the trend in NASA-USAF relations that developed over the space study program was discouraging. The man-in-space project had already been transferred to NASA, and it looked as though NASA would also take over the lunar exploration and base projects as well, with not so much as acknowledgement of indebtedness.

Headquarters USAF, however, made no compromise in its effort to better relations as a long-term objective. In the spring of 1960 it was Air Force policy, and of course ARDC policy as well, to adhere to full cooperation with NASA even "at the risk of our own programs. NASA must have the maximum possible access to ARDC's objectives and aims in projects of mutual interest."[31]

X. ARPA AND THE MILITARY PROGRAM, OCTOBER 1958 TO JULY 1959

The loss of Category IV and most of Category III projects, gave ARPA's program the lean look of starvation. (See program, p 163.) One glance and every bone in the skeleton was visible, but the shrinkage did not alter the agency's position within the Department of Defense. In October 1958 the agency still operated under the original DOD directive of 7 February--whose number had been changed to 3200.5 on 17 May. If ARPA was criticized, sometimes with the audible hope that it might be short lived, Secretary McElroy's reply was always the same: ARPA was a permanent addition.

In the course of the next eight or nine months the overall situation changed. ARPA's program suffered a few more losses to NASA but generally held its own, moved toward maturity in some projects, and gained several new projects. The criticism of ARPA became more and more outspoken in military circles, but the agency was also stoutly defended, and these opinions were personal rather than official statements on the part of the Services. At the same time the position of ARPA shifted and slipped lower in the DOD organizational structure.

The Military Space Program--Second Phase

After the division of projects with NASA, ARPA still did work in the ballistic missile defense area, which in one instance overlapped the space program, and in the space program itself the agency continued active in research and development for various satellites--six in late 1958 and early 1959. Also, ARPA was instrumental in preparing the way for an effective tracking system and supported booster developments as long as permitted to do so.

ARPA's Program

In Summer of 1958 before NASA was Activated

1. Missile defense except Nike-Zeus and BMEWS

2. Military reconnaissance satellite WS-117L
   - 2.1 advanced reconnaissance satellite
   - 2.2 photo capsule
   - 2.3 24-hr reconnaissance
   - 2.4 manned strategic station
   - 2.5 strategic communication satellite
   - 2.6 global surveillance

3. Military developments for the application of space technology
   - 3.1 man-in-space
   - 3.2 special engines
   - 3.3 special components
     - 3.3.1 chemical batteries
     - 3.3.2 nuclear reactor
     - 3.3.3 solar batteries
     - 3.3.4 telemetry (etc.)
   - 3.4 Project Argus
   - 3.5 satellite tracking (Space Track)
   - 3.6 practical application of satellites
     - 3.6.1 communications satellite
     - 3.6.2 meteorological satellite
     - 3.6.3 navigational satellite
   - 3.7 bomb-powered rocket
   - 3.8 solid propellants

4. Other advanced research projects
   - 4.1 ABMA/JPL program
     - 4.1.1 NACA balloon for density
     - 4.1.2 ABMA scientific satellite
     - 4.1.3 Army lunar probe
     - 4.1.4 Army lunar probe
   - 4.2 Three USAF lunar probes (one fired)
   - 4.3 NRL gadget program to photograph back side of moon
   - 4.4 Follow-on program, a continuation of the IGY work in
     - 4.4.1 cosmic measurements
     - 4.4.2 astronomical measurements
     - 4.4.3 solar research
     - 4.4.4. biological research

Probable Program after Transfers to NASA

1. Missile defense except Nike-Zeus and BMEWS

2. Military reconnaissance satellite WS-117L (being reorganized into three separate projects within ARPA)

3. Certain to be lost to NASA except for 3.5 satellite tracking and 3.6.3 navigational satellite

4. Certain to be lost to NASA

The Satellite Projects

Three of the six satellite types being sponsored by ARPA came from WS-117L. On 10 September 1958, Johnson redefined the Advanced Reconnaissance System and broke it down into separate projects with different designations. Previously the system designation had been changed from Pied Piper to Sentry, and now Johnson kept his name for the true reconnaissance satellite that employed visual (photographic) and ferret (electromagnetic) methods of observation. He stripped away a series of experiments that had clustered around Sentry and gathered them together as the function of another satellite, Discoverer. This project was designed for vehicle and subsystem tests, biomedical flights, and the mastery of recovery techniques. The infrared subsystem of ARS then was redesignated as the Missile Defense Alarm Satellite (Midas). Its function was to detect ICBM's at practically the instant of their launching and thereby appreciably advance the time of warning. All three projects were assigned to ARDC-AFBMD with the usual contractual arrangement. These three satellites would depend initially upon Thor boosters but eventually the operational versions would employ Atlas.[1]

A satellite strategic communication station had also been one of the subsystems of WS-117L, but it was not until July 1958 that ARPA acted to support the idea and instructed ARDC to prepare the plan. By that time the Army and Navy too had submitted their communication satellite requirements, and a strong triservice interest was vested in the outcome.[2]

The ARDC abbreviated development plan was completed on 26 August. It called for a worldwide communication system consisting of several satellites in polar orbit and, later, four on equatorial orbit which, at altitudes of 22,000 nautical miles, would equal the angular velocity of the earth and appear stationary. The primary purpose of the system was to alert the United States in a crisis of imminent hostilities, provide SAC sufficient warning to mount a retaliatory strike force, and enable SAC to exercise command and control over the strike force once it was airborne. The Air Staff approved ARDC's plan and submitted it to ARPA on 30 September 1958.[3]

After cautious consideration Johnson approved the plan on 22 October but splintered the project, assigning vehicle development to ARDC and payload to the Army Signal Corps. Here, as in almost no other instance, the splintering became a major issue because Air Force interest in a communication satellite was widely different from that of the Army and Navy. These two services wanted the equatorial satellite to improve the trunking system for the transmission of critical intelligence information to and from Europe, the Middle East, and Far East. Such an aid to communication would provide the Army and Navy with the best insurance of command and control for oversea weapons and forces. Since an equatorial, stationary satellite, orbiting at 22,000 miles, was certainly far beyond the capabilities of the United States in 1958, the Army and Navy wanted interim repeater* satellites orbiting the equator at 2,000 miles. The Air Force had little interest in any equatorial satellite, and none in a repeater, for this satellite could cover only that territory between $75°$ S and $75°$ N latitude. The Air Force wanted a polar satellite that would cover areas of the world where SAC forces were flying.[4]

The Air Force protested strongly to Johnson against his decision, and

───────────

*The repeater principle had been used in Project Score, launched 18 December 1958.

he conceded the need for an Army-Air Force working group to insure that the Signal Corps designed a communication package meeting the needs of all three services. In the discussions that followed the Army and Navy were in mutual support, and agreement with the Air Force was difficult. The negotiations continued for weeks, with no settlement in sight. Johnson broke the deadlock by stating on 6 March 1959 that SAC's need for a polar satellite should be met at the earliest practicable date. His decision resulted in an ARPA-sponsored communication satellite program consisting of three major systems: the SAC polar satellite termed Steer that would have its first test flight in 15 months, using an Atlas-Agena vehicle; an interim delay-repeater satellite designated Courier; and a 24-hour global communication satellite named Decree, to be developed sometime in the future.[5]

In the midst of these discussions, the long-dormant suggestion of a navigation satellite rose to the level of ARPA approval. Both the Navy and the Air Force expressed an interest in such a project when making their recommendations to the Stewart Committee in December 1957-January 1958. But the Air Force did not include it in its 25 January proposal, and the project was generally regarded as one primarily supported by the Navy. The purpose of the navigation satellite was to insure an instantaneous all-weather system for determining the position of any point on the globe by passive means. The receiving station, on ship or shore, or in the air, would listen for a radio signal from the satellite as it came over the horizon. The satellite would relay to the receiving station the signal for the Doppler shift, the synchronous time, and the orbital parameters in effect. The information would be sufficient to permit locating the circle of position within 0.4 mile. The project had the name Transit. It was essentially simple, and in expectation of miniaturization, ARPA was thinking at the end of fiscal year 1959 of employing it operationally as a piggyback payload on some prime satellite mission. During the research and development stage, management was split between AFBMD and the Navy's Bureau of Ordnance.[6]

The sixth satellite project was the elaborated version of the cloud cover experiment assigned to ABMA by AO 1, on 19 March 1958. In succeeding months it had been designated the Television Infra-Red Observing Station (Tiros) and was being developed to observe weather conditions in target areas, refueling zones, landing fields, and ocean operating areas. It would be for all intents an extension of weather aircraft operations. Its payload would consist of television cameras and photocells for infrared detection. As the project expanded, the Army Signal Corps and the Air Force Cambridge Research Center had been admitted to participation. It was a promising project, but in January 1959 Johnson informed JCS it would be transferred to NASA.[7]

In briefing JCS, Johnson dipped somewhat into the future. He spoke of a possible satellite for electronic countermeasures, of a space surveillance platform, and of a maneuverable recovery space vehicle (MRS V). The latter would insure a means of attack, defense, and escape, and the ARPA director expressed his confidence that in the end man-in-space would be possible. In this connection he referred to the loss of the man-in-space project to NASA but pointedly remarked that the Air Force's boost-glide Dyna Soar would surpass the capabilities of Mercury. Here, indeed, was potentially a manned space vehicle that could maneuver in and out of orbit, remaining under sufficient pilot control to operate from and return to predetermined fixed military bases.[8]

It was gratifying to have Johnson's recognition of Dyna Soar, but to some USAF officials his remarks could have presaged its loss either to ARPA or to NASA. ARPA saw the need of a manned maneuverable spacecraft and could take over Dyna Soar for development if the Air Force advanced its orbital capability. At the same time, NASA claimed to be the agency for manned spaceflight and could demand the transfer of Dyna Soar if ARPA took it as a manned space vehicle. As one Air Force officer said:[9]

> The Air Force has been successful in retaining control of Dyna Soar by asserting that it has less than an orbital flight capability. This procedure has thus far succeeded in thwarting ARPA's overtures to take over the program. The Director of ARPA has stated that the Dyna Soar program is the best approach toward the goal of manned space vehicles having a military capability. It is anticipated that ARPA will develop some type of a man-in-space program patterned after the Dyna Soar program. At the present time, ARPA is conducting investigative studies on advanced vehicle characteristics which would be applicable to such a program.

As a safeguard, the Air Force continued for some time to emphasize the suborbital rather than the orbital characteristics of Dyna Soar while going forward with its development as rapidly as weak funding and strong opposition within OSD permitted.*[10]

---

*At the time that Johnson briefed JCS, January 1959, Dyna Soar seemed to be moving forward rapidly. On 25 November 1957, DD 94 authorized ARDC to proceed with Dyna Soar, and on 16 June 1958 the command announced the selection of Boeing and Martin as dual contractors for the early design phase. In April 1959, Boeing and Martin submitted designs, and SAB lent full support. Gradually, opposition in OSD seemed also to be dwindling. The Air Force felt more certain of its claims to Dyna Soar and by the late autumn of 1959 was speaking without constraint of the boost-glide vehicle as possibly meeting USAF space requirements. (Draft memo by D/AT, to be sent by SAF to SOD, 23 Oct 59, subj: Required Action on Dyna Soar.)

The Booster Program

As the need for larger satellites became more pressing, the need for more powerful boosters became more and more apparent. In the first six months of 1959, ARPA still had two major propulsion development projects under way--Saturn and Centaur. Saturn was the new name given to Juno V, the 1.5 million-pound booster consisting of a cluster of eight engines. It had begun with in-house studies by ABMA in April 1957, and on 15 August 1958, AO 14-59 directed the Army Ordnance and Missile Command (AOMC) to develop the booster. On 19 May 1959, ARPA announced the selection of the upper stages for the Saturn vehicle--a two-engine Titan for second stage and a Centaur as third stage. The Titan would contribute a thrust of 367,000 pounds, and Centaur would add 30,000 pounds. Before the end of fiscal year 1959, there were prospects of using Saturn to launch USAF strategic surveillance satellites.[11]

Centaur promised great versatility and in 1959 was also being mentioned as the third stage for Atlas or Titan in launching communication and several other kinds of satellites. Its wide usefulness made Centaur applicable to the civilian space program as well as the military, and in February there were rumors--well substantiated--that NASA planned to take over Centaur development at the end of fiscal year 1959. After a long series of negotiations between ARPA and NASA, the former agreed in April to transfer the engine to NASA on 1 July. The shift was a blow to the military who believed the civilian agency would place the project on a slower schedule than that advocated by ARPA and delay its availability for the military projects. Several segments within the Department of Defense continued to argue against the decision, but to no avail. The transfer was effected on schedule.[12]

The Tracking System

The satellite tracking and surveillance system was the most complicated of ARPA's projects in 1958 and 1959. The Air Force originated work on the system in the unhappy days after Sputnik I and II, when it became evident that the launching and flight of friendly and hostile space objects required some form of monitoring. From the viewpoint of national security, it was essential to detect, identify, and if possible, determine the purpose of any satellite.[13]

For want of better equipment, the Air Force turned to radar facilities being developed to detect ICBM's. On 5 October 1957 the Millstone Hill radar at Westford, Mass., prototype of stations to be used in the Ballistic Missile Early Warning System (BMEWS), became operational. Because it successfully tracked Sputnik, the station was believed the best available means of tracking satellites and, along with facilities in Trinidad, became the foundation in the slow buildup of a satellite tracking system.

On 3 December 1957, Headquarters USAF gave primary responsibility to ARDC for coordinating satellite data from radio, radar, optical, and photographic coverage. ARDC, in turn, would transfer the data to the Naval Research Laboratory (NRL) and the Smithsonian Astrophysical Observatory. This arrangement was the first move toward creating national procedures for tracking space vehicles. Plans progressed rapidly and in January 1958, Project Space Track got under way with ARDC's establishment of a filter center at Air Force Cambridge Research Center.[14] Shortly afterwards, six other ARDC centers received contributory assignments. By early April 1958 the project was moving along smoothly. This success, and the growing space program under ARPA, increased the complexities and requirements of the project. The emerging plans for boost-glide vehicles and for reconnaissance, communication, navigation, weather, and scientific satellites, as well as hopes for lunar and interplanetary probes, meant a rapid increase in celestial traffic. On 19 June 1958, Headquarters USAF issued GOR 170, Satellite Defense System, setting forth operational requirements for a tracking and control system.

Meanwhile, the original Air Force-sponsored project was being expanded by higher authority into a more comprehensive program. On 18 January 1958, Holaday, who was still the Special Assistant to the Secretary of Defense for Guided Missiles, directed the Secretary of the Navy to work for the integration of all DOD tracking and surveillance agencies into a national capability.[15] Technically, the chief problem was that of dealing with the "dark" or nonradiating bodies. In attempted solutions, NRL set up a triservice committee which advised OSD that ARPA should establish a "surveillance fence" across the United States. This could be done by utilizing Army and Navy Doploc and Minitrack detection facilities already in existence or under development between San Diego, Calif., and Ft. Stewart, Ga. On 20 June 1958, ARPA directed the Army and Navy to combine their facilities into a fence along the southern border of the country.[16]

The establishment of the fence created yet another problem--the need to have an organization to operate the system. The Army and Navy wanted a new triservice agency; the Air Force wanted responsibility assigned to the North American Air Defense Command (NORAD).[17] Discussion continued through the summer, and since agreement proved impossible, ARPA established the Space Surveillance Task Force. It was composed of representatives from the entire intelligence community. Its purpose was to study the problem from the viewpoint of intelligence requirements as modified by available techniques. The task force recognized the capabilities of the Air Force intelligence agency, the USAF Space Track project, and the filter center and pointed out that the Air Force had already solved the problem within its own service framework. The next step was to place the work on a national basis.[18]

At this point, the President signed the Space Act of 1958. No one knew what NASA's attitude would be toward surveillance. The Air Force hoped for cooperation, but Dr. Glennan promptly asked ARPA to transfer the entire responsibility for developing the detection and tracking unit. Johnson refused on the ground that military interests would be injured.[19]

On 5 November 1958, Johnson informed ARDC that the time had come for an interim control system. Johnson felt that the logical place was the Air Force Cambridge Research Center and directed ARDC to develop the required control units. The data readout facilities of NRL and the Army's Ballistic Research Laboratory (BRL) at Aberdeen, Md., would be made available as necessary.[20] Johnson's decision in a sense acknowledged the great progress made between 18 January and 5 November 1958 by DOD to create a national space surveillance program, and his choice of the Cambridge center as the site for the interim system pleased the Air Force. It did not please the Army, Navy, or NASA.

Before the end of the month, ARDC set up a steering committee to work with representatives from ARPA, USAF, NRL, and BRL. At the meetings, the Army and Navy representatives were openly critical of Johnson's approach. Out of patience with tedious objectives, Johnson directed ARDC, on 19 December, to proceed with the program, and on 13 January 1959 issued ARPA's

System Development Plan for Space Track. It called for an Interim Space Surveillance System (ISSS) to be operational by March 1960, consisting of a worldwide net of sensors feeding information to the Cambridge control center for data processing. Later, a formal National Space Surveillance System (NSSS) could be worked out.[21]

Meanwhile, following the NASA request of October 1958 for complete control of detection and tracking, negotiations had been under way to devise a method whereby the requirements of the civilian agency and the military could be met. A DOD-NASA agreement was signed on 10 January 1959, recognizing the differences between the civilian and military needs. NASA was primarily interested in information on flights pertaining to research and development; ARPA was primarily interested in operational flights that would be of significance for the intelligence community. It was therefore agreed that NASA and ARPA would both operate detection and tracking stations with complete exchange of information. NASA assumed responsibility for a three-station net, with stations in California, Australia, and South Africa, and some Minitrack stations for polar and Mercury flights. The Department of Defense was then left with responsibility for detection and tracking from the Atlantic, Pacific, and White Sands missile ranges, the east-west Minitrack fence, and two additional stations--one in Japan and one in Spain.

Management of the national system was entrusted to an ARPA-NASA technical committee organized without service representatives. Moreover, the committee worked directly with ARDC and Cambridge, bypassing Headquarters USAF. The situation was far from satisfactory as far as the Air Force was concerned. Nevertheless, great progress had been made toward an effective NSSS in which the Army and Navy were responsible for creating a more effective method of detecting and tracking noncooperating satellites and the Air Force provided additional sensors and the control center.[22]

At the End of June 1959

On 17 March 1959, OSD canceled DOD Directive 3200.5 under which ARPA had operated since 17 May 1958, and issued a new DOD Directive 5105.15. Though it revived the number of the original directive of 7 February 1958, it actually "wiped the slate clean" of the many changes of the past year. It effected a redefinition of ARPA's program in a series of descriptive inclosures, which grew to more than a dozen between March and July 1959.

Under the new directive, ARPA continued to operate in three areas, though two of them were somewhat broadened beyond what they had been. Missile defense was now termed defense against "extra atmosphere offensive vehicles," to include both space vehicles and ballistic missiles; and propellant chemistry seem now in close alliance with vehicle materials. The area of the military space program remained the same, but within it there was considerable shift in projects.

The titles of the projects assigned to ARPA under the new directive reflected the progress being made in the space program and its related fields. Project Defender covered the activities of ARPA in devising methods of defense against hostile missiles, satellites, and other space vehicles. Project Principia and Project Pontus pertained to propellant chemistry and vehicle materials.

In the space program itself, there were 10 projects. Discoverer, Midas, and Sentry* were elements of the old WS-117L ARS, now separated as the test experiment, infrared detection, and reconnaissance satellites

---

*Sentry was to be redesignated as Samos in August 1959.

respectively; Notus and Transit were the communication and navigation satellites; Sheppard was the space surveillance project; and Suzano was a more recently approved space platform or orbital base from which to launch advanced space missions. OSD had also approved a space electronic countermeasure project dubbed Somnium; added a requirement for Tribe, a series of vehicles for special military space missions; and established Project Longsight, a series of space studies and system analyses to supply DOD on a continuous basis with suggestions of projects that should be initiated to satisfy future military requirements. There was a strong similarity between the stated purpose of Project Longsight and ARDC's space study program.

ARPA's Changing Status in DOD

For months in 1958, ARPA held a unique and powerful position in the Department of Defense. The director was the voice of the Secretary of Defense in matters of space research and development. Sometimes called a "fourth service" or a DOD "task force," the agency had seemed--to JCS and the services--more like an arm of OSD reaching down into the operational level of the military services.

The Army, Navy, and Air Force felt from the beginning that ARPA should not be permanent, but as time passed it was evident that ARPA was doing an excellent job, and the opposition of the Army and Navy waned. They still agreed with the Air Force in late 1958 and early 1959 that ARPA should go, but they did not agree it should go soon. Navy spokesmen, for instance, said that the Navy was not in the space vehicle business and was interested much more in payloads, "in what goes into space," than in how the payloads would get there. The implication was that the Navy was no longer disturbed by ARPA as a fourth service since the agency had not attempted to interfere with matters pertaining strictly to the sea.²³

The Air Force, on the other hand, was very much in the space vehicle business since its missiles were being used as boosters for so many satellite and space probe launchings. General Schriever was glad to acknowledge the good work done by ARPA but voiced sharp criticism of its disregard of tested concepts of management, its practice of splintering projects among the services, and its failure to recognize the urgency of defining a military posture in space. In January 1959, Schriever did not hesitate to say that ARPA should be phased out at the end of fiscal year 1959. The elimination of ARPA would leave DDR&E to become the space policy agency and permit the services to do their own research and development as they had done for land, sea, and air requirements in accordance with definite mission assignments.²⁴

After Dr. Herbert F. York's appointment as the first Director of Defense Research and Engineering on 24 December 1958,* there was confusion as to whether Johnson outranked York or York outranked Johnson. The situation elicited some amused but unfavorable comment in congressional hearings.²⁵

The question was not settled until 17 March 1959, when DOD Directive 5105.15 enumerated ARPA projects and went on to say that they were "subject to the supervision and coordination of the Director of Defense Research and Engineering in the same manner as those of the military departments and will be conducted in accordance with the priorities established by the

---

*See above, p 109.

Secretary of Defense." The directive thus slipped a new echelon between the Secretary of Defense and ARPA.

In May 1959, York explained the arrangement in the course of testimony before a congressional committee. He said there were four basic operating agencies in the Department of Defense--the Army, Navy, Air Force, and ARPA--doing research and development either by in-house work or by contract with outside agencies. DDR&E would supervise and coordinate all research and development, including that assigned to ARPA.[26]

The changes brought about by the establishment of DDR&E were far short of what the extreme critics of ARPA would have liked. But, whether they meant a deliberate change in the attitude of OSD toward ARPA or not, they certainly deprived the agency's director of some of the authority given him in February 1958. He was no longer the voice of the Secretary of Defense in matters of space research and development. That authority was now vested in York.

Space Operations, October 1957 to July 1959

In the first 20 months of space operations, the Russians made four successful launchings. They admitted no failures. The Americans on the other hand attempted 26 launchings--21 earth satellites, 3 lunar probes, and 2 interplanetary probes. Eight satellites entered orbit and 13 failed; the three lunar probes failed, not exceeding 70,000 feet, 71,000 miles, and 1,000 miles, respectively. One of the interplanetary probes reached an altitude of 63,582 miles, and the other went into orbit around the sun.

This record put the Americans far ahead of the Russians if numbers alone counted. However, there were other factors to be considered. The Russians payload began with 182 pounds for Sputnik I and increased in

Sputnik II to 1,120 pounds. Furthermore, the Russian launchings led to important "firsts"--first to send a satellite in orbit around the earth; first to send animal life in orbit around the earth; first to have an interplanetary probe place a satellite around the sun. The American weakness was lack of thrust to send up large payloads, and the United States started with a payload of 3.5 pounds and increased the weight eventually to a maximum of 372 pounds.[27]

The Russian triumphs had great psychological-political significance. Dr. Glennan admitted as much when he said in September 1959 that Americans still "play second fiddle in this space buisiness." The President and the National Security Council expressed the same view officially and explicitly, but not publicly, when they acknowledged in January 1960 that the Russian "firsts" resulted in "substantial and enduring gains in the Soviet prestige."[28]

Nevertheless, thanks to the ingenuity and devotion of NASA, ARPA, Army, Navy, Air Force, and industrial scientists, the Americans made noteworthy contributions to space science through the use of miniaturized instruments. These were made specially to compensate for the nation's lack in rocket-engine thrust power. The eight orbited satellites were distributed among four projects--2 Vanguards, 3 Explorers, 1 Score, and 2 Discoverers.

Vanguard alone had received official approval before Sputnik and was intended to serve solely as a scientific contribution to IGY. Explorer was hastily conceived, primarily as a countermeasure to the Russian Sputnik success. Incidentally, it too served to gather IGY scientific information though it depended for propulsion upon the use of a military missile as the lift device. Project Score was the Christmas greetings satellite successfully launched on 18 December 1958.

Of the four projects, Discoverer was the first true military satellite. Painstakingly prepared by AFBMD after being separated by ARPA from WS-117L, it was instrumented to aid in the development of other military satellites. It had six main test objectives: airframe and guidance subsystems, stabilization equipment, means of controlling the internal environment, reaction of mice and small primates to weightlessness, adequacy of capsule recovery techniques, and proficiency of ground support equipment and personnel.[29]

Discoverer I and Discoverer II were successfully launched on 28 February and 13 April 1959. Discoverer II created a small international incident of rivalry, intrigue, and theft. The satellite was the first to contain a recovery capsule, equipped with a retro-rocket ejection mechanism. It was intended to permit recovery after a few passes around the earth, and the recovery task force, consisting of nine C-119's, four RC-121's, and three destroyers, was operating off Hawaii. Several attempts to trigger the capsule failed. Not until the next day, during its seventeenth pass, was the capsule ejected--and this was automatic. It came down reputedly in a fjord northwest of Longyear City in the Spitsbergen Islands. The islands were Norwegian, but under a provision of the 1920 Treaty of Paris, all signatory nations, including Russia, had the right to exploit mineral deposits there. Numerous Soviet mines existed in the region and the island was inhabited by Russians in 1959. The capsule, therefore, fell almost in Soviet laps. It was not surprising that neither the Americans nor the Norwegians could find the capsule, but they saw evidence that the Russians had found and shipped the prize home to their own space scientists. The

April issue of Current Intelligence Digest published the following comments on the incident:

>A search party was immediately organized and the search continued through 22 April with negative results. During the search in the Spitzbergen area, a helicopter crew observed footprints in the snow and evidence of a heavy object being dragged through the snow into the entrance of a Soviet mine. The Air Attache in Oslo, Norway, reported that discussions with Colonel Tatum, Air Rescue Commander, and Lt Colonel Metheson, AFBMD, revealed that Russian indigenous personnel did retrieve the capsule on 15 April. The USAIRA, Oslo, also reported that several days later, a Russian ice breaker was sighted entering and departing Longyear City, Spitzbergen.

XI. AIR FORCE SPACE POLICY AND SUPPORTING ACTION, 1957-59

Perhaps the most disturbing thing about Sputnik was the paradox of its undeniable importance and its imprecise significance. There was universal pride in the satellite and an intuitive understanding that history was in labor with one of its great crises. Change was imminent, but without clairvoyance of what the change would be, the world strained with anxiety for the future--and with nostalgia for the comfortable ways of Earth.[1]

In the conflict between the opponents and proponents of a military space program, no one denied that a scientific program was essential because little was known about space; but for that same reason many denied the need for a military program. To this argument the military replied that despite existing ignorance of what space warfare would be, space, as an area of operations, would eventually shatter the old-time concept of land, sea, and air missions. In time, space weapons would erase existing lines of responsibility. The United States must develop a military space competency parallel with scientific knowledge or one day find itself helpless under a devastating blow. The question was whether the nation should seek merely the scientific exploration of space or both the exploration and control of space. The military were careful to point out that capability to control space did not necessarily mean the abusive exercise of that control.[2]

In the late months of 1957 and early 1958, the military frankly admitted they could speak less fluently of space warfare than of infantry maneuvers, of naval blockades, or of strategic bombing. They lacked the experience. "There is a lot of scientific data that we can get from exploration of space," said General Schriever in April 1958, but it is "impossible, I would say now, for any man to predict exactly how important space will be for military purposes, looking into the future, 20, 30, or 40 years."[3] In December 1959, an Air Force position paper commented:[4]

> Scientific and engineering contributions to the solution of military problems have revolutionized warfare. The invention of gunpowder, the steam engine, the submarine, the airplane, the tank, radar, nuclear weapons and the ballistic missile, to name only a few examples, have had profound effect on military strategy and the balance of power between nations.
>
> The latest contribution and perhaps the greatest technological change of all, is man's first step into space. It can be clearly foreseen that military space systems will alter current military concepts and strategies, even though the exact nature of their use and effect cannot be delineated over the long range. Recognizing that the military potential of space is of such significance that within this century it may well determine the future history of the world, the military must exploit this potential in the national interest.

If the military did not succeed in the two years since October 1957 in reaching a clear conception of space warfare, it was not for want of trying. The effort was persistent, but the objective was too big to be seen without perspective. Army, Navy, Air Force, and JCS analysts attempted time and again to foresee the role of the military in space but ended always with nothing more than a laborious listing of satellite projects under development. A list of ship classes could not explain naval strategy or tactics, or the significance of sea power, and lists of possible space vehicles did not explain space strategy or tactics, or the significance of space power. The best that could be done was to speak of the "military uses of space."[5]

The inability of strategists and tacticians to prepare handbooks on space warfare in no way detracted from the obviously great military significance of space. The Army, Navy, and Air Force knew that somehow, and in some way, space would become the transcendent factor in preserving peace or, failing that, in winning the war. That knowledge was the important thing, and the DOD space projects were as much exploratory for military ends as the scientific projects were exploratory for scientific ends.

In the first two years of their exploratory efforts the military were keenly aware of the need to reconcile their requirements with a national space policy that leaned more favorably toward the civilian-scientific program. The services had also to fortify their separate positions with doctrine and with attempts to obtain assigned roles and missions. The prospects of space were limitless, and the services vied among themselves for dominance.[6]

Military Reactions to the Space-for-Peace Policy

After January 1948, when General Vandenberg asserted that responsibility for satellites logically belonged to the Air Force, there was no further policy statement on space prior to Sputnik. In the confusion that followed Sputnik, the Air Force felt the need of a new statement, especially as a talking paper for its representatives appearing before congressional committees. The task of preparation fell to the Office of the Deputy Chief of Staff/Development.

On 6 December 1957, DCS/Development forwarded to the Chief of Staff, USAF, a policy statement that affirmed the loyalty of the Air Force to national objectives and asserted that the control of space was essential to national security. Continuing, the paper declared that the Air Force was the logical agency to achieve this military power because there could be "no division, per se, between air and space; only one indivisible field of operations above the surface of the Earth."[7]

In referring to Air Force support of national objectives and the desire to exploit the military advantages of space, the policy statement fingered an unpleasant dichotomy from which the military could not escape during the next two years. The national objective was expressed in the President's space-for-peace policy, so phrased in pre-Sputnik days as to exclude the military from those regions beyond the aerodynamic capabilities of airpower.* Had this same principle been part of the freedom of the seas, the navies of the world would have been excluded from the oceans beyond the three-mile limit. None denied the ideal of space-for-peace but the resultant restrictions did not jibe with the obligations of the military for the security of the United States. From the beginning to the end, the Army, Navy, and Air Force were united in expressing their acceptance of space-for-peace as the national objective, but until international arrangements could guarantee that all nations would follow the same ideal, an American capability to control space was essential to the liberty of free people.[8]

---

*Even before Sputnik the Air Force was aware that space had serious implication for airpower. In the spring of 1957 the Air Force General Counsel undertook a study of "air space" and "outer space" definitions, the legality of satellites in transterritorial flight, and the legal complications of space defense. Of equal interest were the less cosmic subjects of interservice rivalry, military budgets, and possible changes in the structure of the Air Force. The General Counsel moved slowly, however, and in March 1958 was still weighing the advantages of various alternative policies before recommending an official position. (Memo, Col R.L. Johnson, Dev Div, D/R&D to D/Plans, 15 May 57, subj: Future of Satellite Operations.)

In seeking to adjust their loyalty to the President's somewhat extreme position and their obligations to safeguard the defense of the United States, the military did not criticize the space-for-peace policy but sought rather to determine for themselves how effective international space law was likely to be, how it would curtail their own activities, and how far they should go in presenting a case for military space projects.

In March 1958 the three services took advantage of their invitation to assist the National Security Council in its preparation of the Preliminary United States Policy on Space (NSC 5814/1). They expressed their views and were at one in supporting the ideal of space-for-peace. They were also at one in warning against emasculating the military program. After its NSC recommendations, the Air Force undertook a second study, for Air Force eyes only, on the feasibility of an international law for space and its effects on the military space program. The task required five months to complete, and DCS/Plans and Programs drew upon the advice of all interested Headquarters agencies as well as the Air University and Rand Corporation.[9]

On 22 August 1958 the Air Force "Study on Sovereignty over Outer Space" was completed, and by 8 October it was distributed among Headquarters offices with the recommendation that it serve as the basis for developing future studies by the Air Staff when called upon for comments or actions related to an international law for space. It was Air Force doctrine, under this paper, that the Government should avoid committing itself on any current issue. Time was needed to evaluate the totally new conditions being created by the space age. It seemed particularly unfortunate for the Department of State to assume, as it was assuming, that silence on space claims in relation to specific events, such as Sputnik's transit, implied a general waiver of claims. Effective international control of space conceivably could come in the future, but it was a goal not a reality. In supporting the President's policy, the military should urge and assist in obtaining international cooperation in projects not pertinent to national security, thereby contributing to the eventual attainment of the national objective. At the same time the Air Force should seek approval of an adequate program of research and development and work toward the formulation of projects to meet the scientific, commercial, and military needs of the United States. The military goal must always be the prevention of Soviet dominance in space.[10]

The passing months brought no variation in themes. The Government continued to speak of space-for-peace, to negotiate in the United Nations for acceptance of the policy, to find powerful support for the idea in Congress and among the people generally. The military continued to face the dilemma with the same uneasiness. On 2 May 1959 the Chief of Naval Operations wrote to the Chairman of the Joint Chiefs of Staff:[11]

> I have noted recent statements by members of Congress, and of the Executive Department, with a generally favorable and public reaction on "the peaceful use of outer space." I view with concern the adverse effects on national security of an unrealistic or restrictive international agreement to use outer space solely for peaceful purposes. The U.S. Military Services have responsibilities which require the use of outer space for research, development, and operation of weapons systems. Until an enforceable and positive guarantee of control in the use of space can be made, this will remain so.

The Joint Chiefs on 22 May forwarded an elaboration of these views to the Secretary of Defense. It would be, they said, a most serious matter to restrict the services in their use of space before an enforceable international control came into being. The current and future capabilities of the armed forces should not be hampered by premature agreements to keep space for peace, and the JCS hoped that its opinions would be considered before the United States committed itself to definite agreements.[12] Seven months later, in December 1959, the Assistant Vice Chief of Staff, USAF, Maj. Gen. Richard M. Montgomery, said essentially the same thing:[13]

> The President's announced policy is /that/ the exploration of space and, therefore, this country's space programs, will be used only for peaceful purposes, and for the good of mankind. Failing that, however, the Air Force believes that there is a great potential in space from a military standpoint, and that this potential must be developed.

The Doctrine of Aerospace

The service chiefs were not always in unanimous agreement on questions of space. Nothing divided them more sharply than the USAF claim that the continuum of air and space gave the Air Force the responsibility, under accepted roles and missions, to become the service of primary interest in space. This had been the argument, implicitly at least, in Vandenberg's statement of January 1948. It was repeated in the policy statement of 6 December 1957. And again, in March 1958, the Chief of Staff, USAF, General White, wrote:[14]

> For all practical purposes air and space merge, form a continuous and indivisible field of operations. Just as in the past, when our capability to control the air permitted our freedom of movement in the land and seas beneath, so, in the future, will the capability to control space permit our freedom of movement on the surface of the earth and through the atmosphere.

Though numerous Air Force officers repeated the same thought in public statements, in articles, and in classified papers during the succeeding months, the doctrine of air-space continuum was not propagated systematically. By the time 1958 was well advanced, ARPA and NASA acquisitions and actions had virtually wrecked the proposed USAF astronautics program of

25 January, and the Air Staff felt it was essential to undertake a counteroffensive to regain if possible some portions at least of its program. There was need to reassert the doctrine of air-space continuum and define it in terms that lent emphasis to the concept. The movement came to a climax in the last days of 1958 when the Air Staff was preparing a talking paper for the Chief of Staff in his scheduled appearances before congressional committees. This meant redoing the policy paper of 6 December 1957 with emphasis on what the Air Force had contributed to both the military and scientific space programs during 1958, a reassertion of the air-space continuum doctrine, and a restatement of Air Force claims to overall responsibility for the development and control of space vehicles, granting recognition of limited Army and Navy needs. The policy statement was completed on 30 January 1959 and was consonant with the accepted roles and missions for operations on land and sea and in the air. It spoke of the air-space continuum as "aerospace"--a term used for some time within a few Air Staff offices--and defined the term in such manner as to justify Air Force claims as the service of primary interest there.[15]

On 3 February 1959, General White appeared before the House Committee on Science and Astronautics. He expressed the desire of the Air Force to promote the peaceful use of space for the benefit of mankind as sought by the President. Pending effective measure o that end, however, White declared that the right of the Free World to explore space depended upon a "strong and capable deterrent aerospace force." He added:[16]

> The Air Force has operated throughout its relatively short history in the sensible atmosphere around the earth. Recent developments have allowed us to extend our operations further away from the earth, approaching the environment popularly referred to as space. Since there is no dividing line, no natural barrier separating these two areas, there can be no operational boundry between them. Thus air and space comprise a single continuous operational field in which the Air Force must continue to function. The area is aerospace.

General White thus brought to the fore the old doctrine of air-space continuum.

The significance of the definition was not missed, and the reaction was less than unanimously enthusiastic. The criticism that rose within the committee overflowed and spread far beyond the limits of the room. The next day Representative Daniel J. Flood of Pennsylvania, in hearings before a subcommittee of the House Committee on Appropriations, allowed himself the pleasure of sarcasm. He epitomized the feelings of many Air Force critics, both civilian and military in the following dialogue:[17]

Rep. Daniel Flood:

> Boys, the Air Force has come up with a new phrase, "aerospace". That is a beauty. Even Winchell could not think that one up. That means everybody is out of space and the air except the Air Force, in case you don't know it. Has the Air Force, without consulting anybody, taken the Navy out of the air and space? . . . /The Air Force has/ now staked out a claim to "aerospace". That is their pigeon--space and air. Do you know about it /Mr. Gates/? . . . You Navy people had better get into this space thing, because I have been around here for a long time, and I have seen this happen in other areas. You had better get back into space or you "aint" going to be in space.

Thomas S. Gates, Secretary of the Navy:

> We have a little thing called Vanguard which is doing pretty well.

Rep. Daniel Flood:

> I know, I agree with you. But the honeymoon is over for the Air Force. There will not be many aircraft around when the sons of the flyboys go to the Air Academy. They have to have something to stay in business. You had better get into there, or you won't be around.

General White did not retract the claim of the Air Force to aerospace. The criticism at least accustomed all ears to the sound of the word, and it made known the position of the Air Staff. The discussion continued through the remainder of 1959 but gradually the comments lost sharpness. In its final form, officially approved and incorporated in AFM 1-2,

December 1959, the definition stated:

> The aerospace is an operationally indivisible medium consisting of the total expanse beyond the earth's surface. The forces of the Air Force comprise a family of operating systems--air systems, ballistic missiles, and space vehicle systems. These are the fundamental aerospace forces of the nation.

Logically the doctrine of aerospace expressed the thought that airpower and space power are the same thing and should be vested in a single service which, whatever its official title, would be the aerospace force. Space vehicles would be another category of vehicles for employment in the regions above the surface of the earth to help deter war or, failing that, to help win the war.[18]

The Air Staff felt that the doctrine of aerospace epitomized USAF space policy, completely adjusted to the accepted theory of roles and missions. But a doctrinal pronouncement did not automatically correct the situation. Historically and logically the Air Force claims could be justified, and they could even be made to fit the Air Force needs and capabilities as projected into the 1960-70 decade--though prophecy was dangerous in an age of technological change. Nevertheless, the unhappy fact remained that in the first part of 1959 the Air Force still had no space program in being that could justify claims to space leadership.[19]

The task confronting the Air Staff was to devise some way of persuading OSD to assign space missions and projects to the Air Force to fill the vacuum created in 1958 by the losses to ARPA and NASA. There were two possible approaches: to request a redefinition of service roles and missions not yet updated to meet the demands of the space age* or to pursue a slower

---

*At the end of 1958 the functional responsibility of the Air Force derived from three documents: the 1948 Key West functions paper, DOD Directive 5100.1 of 16 Mar 1954, and its revision of 31 Dec 1958. The Key West Agreement was prepared without thought of the space age, and no changes had been introduced. Nor were there provisions in 5100.1 to meet the current technological developments or the new political-military situation. (Ltr, Gen T.S. Power to Gen T.D. White, 9 Feb 59.)

course but accelerate development of specific aerospace hardware with the approval of ARPA and NASA. To request clarification of missions without possessing the hardware might provide the Army and Navy an excellent opportunity for refutation to the permanent detriment of Air Force claims.[20] Headquarters USAF chose to move slowly, and General LeMay, Vice Chief, explained this approach in a letter of 17 March 1959 to General Power at SAC:[21]

> Specifically, we intend to accelerate the development of aerospace hardware and intensify our efforts to obtain early official sanction from the National Aeronautics and Space Administration and the Secretary of Defense for the Air Force to pursue these development projects . . . . While recognizing Army and Navy interest in aerospace projects, we would seek to limit their participation to a coordinating role. Furthermore, we are making every effort to place qualified Air Force representatives in key positions of influence in the Office of the Secretary of Defense and other governmental organizations concerned with space activities.

The Navy-Air Force PMR Disagreement

In 1957 and the early part of 1958 the Army seemed to have better space claims than the Navy. There was the undisputed fact that the Army could have put a satellite in orbit before Vanguard; and there were the impressive Explorer orbitings on 31 January, 26 March, and 26 July 1958. Slowly the tide turned. There was the launching of Vanguard I on 17 March 1958 and the USAF success in placing the Project Score satellite (Atlas missile) in orbit on 18 December 1958. Thus, the Air Force and Navy both showed a competence to equal that of the Army before the beginning of 1959. If it had not been for the Saturn booster project, the Army would have had an insignificant space role for the future. Even with Saturn, the trend was toward a Navy-Air Force race.

In the spring of 1959 the Chief of Naval Operations established an ad hoc committee under Rear Adm. Thomas F. Connolly to determine the

Navy's astronautics policy. Three months later, CNO approved the Connolly report. It called for a comprehensive program to enhance the "roles and missions presently assigned to the Navy" and cited the Pacific Missile Range (PMR) as a major contribution by the Navy to the national space program. At once CNO established in his office the Astronautics Operations Division (Op-54) and the Astronautics Development Division (Op-76). At the same time the Navy assumed a more aggressive attitude in the long-standing dispute with the Air Force over PMR-Vandenberg AFB relations.[22]

The Pacific Missile Range and Vandenberg were in juxtaposition because both were carved from an 86,000-acre strip of California coastland, formerly the Army's Camp Cooke. In 1956 the Air Force obtained 67,000 acres of the northern part of the tract as a training-operational area for ballistic missiles. It was named Cooke AFB until redesignated for Vandenberg in October 1958, and it was placed under the administrative and operational control of AFBMD's 1st Missile Division (later reassigned to SAC). Before the end of 1957 the Air Force was constructing Thor and Atlas pads at Vandenberg that could be used either for training or operational purposes. As soon as a military space program seemed probable, it was evident that Vandenberg offered special advantages for launching polar-orbiting satellites, since the boosters would pass over no land north of Antartica.[23]

Meanwhile the Navy obtained permission from OSD in December 1957 to combine the remaining southern part of Camp Cooke, totaling 19,000 acres, with the Naval Missile Test Center, which had been operating at Point Mugu, 90 miles to the south, since 1946. The Navy would develop its new possession, the Pacific Missile Range, as a national facility and companion to the Atlantic and White Sands missile ranges.[24]

Since PMR would serve all agencies needing its special facilities, the Navy and Air Force were certain to overlap in operational activities and interests. To forestall misunderstandings, Adm. Arleigh Burke, Chief of Naval Operations, and General White, Chief of Staff, USAF, began negotiating early in 1958 for an agreement, which they signed 5 March. It provided for coordination in fixing radio frequencies, firing schedules, and the avoidance of undesirable duplication. Range safety remained with the Navy or the Air Force depending upon the responsibility for tracking specific missiles, but this provision was in contradiction with the Navy's proprietary authority for flight preparation of missiles, control through flight and impact, and operation of the range safety equipment.[25]

The agreement was of little benefit. The Navy formally opened the Pacific Missile Range in June 1958, and at once Navy-Air Force misunderstandings began. The trouble started at the operational levels of PMR and the 1st Missile Division. Small questions, such as basing a Navy drone aircraft on Vandenberg, were followed by accusations and counteraccusations of improper coordination and lack of cooperation. Other contentions involved the means of insuring safety for Southern Pacific Railway trains passing through PMR-Vandenberg territories during missile-satellite launchings, the authority to reimburse the railway company for necessary interruption of its schedules, and the right of PMR to negotiate a unilateral agreement with the railway regardless of objections either by the 1st Missile Division, SAC, or Headquarters USAF.[26]

A much more serious misunderstanding arose in February 1959 when ARPA directed the Air Force to launch Discoverer satellites on 28 February, 13 April, 3 June, and 25 June into polar orbits from Thor pads at Vandenberg.

The Discoverer trajectories passed over PMR territory. Each time for the sake of safety, the Navy evacuated PMR, where work was proceeding on two Atlas pads for the Samos and Midas projects, and the entire village of Surf, located between PMR and Vandenberg AFB. The Navy also halted Southern Pacific trains. The cost of these evacuations and delayed work, which lasted for several hours or several days depending on the countdown, ran into considerable sums, and of course proved most annoying to the civilians affected. The Air Force felt that the evacuations were unnecessary since it considered the odds to be 200,000 to 1 against fatalities. The Navy's rebuttal was that the odds were only 20,000 to 1. The real point at issue was responsibility for safety, a question complicated by the contradiction implicit in the Burke-White agreement of March 1958.[27]

For the Air Force the significance of the PMR controversy involved not only the control of important facilities on the west coast but also strategy, roles and missions, and the future functions of the Navy and Air Force in space.[28] On 13 December 1958, General Power wrote General White:[29]

> The Navy appears to be using custodianship of the Pacific Missile Range to develop an ambitious space program centering on this range. The implications of Navy planning are not significant on the surface but could be devastating. The precedent for the control of space could be well established, yet easily disguised with the development of world-wide launch, tracking and control facilities. The agency controlling space facilities will control the space missions.

Power cited two documents in support of his contention. A special OSD committee report of 18 August 1958 recommended transfer of the southern part of Vandenberg to PMR, and an Aerojet Corporation report, prepared under contract with PMR, recommended expansion of the range to include six major divisions: Polar Orbit Range, Equatorial Orbit Range, Intercontinental Range, Intermediate Range, Anti-Missile Range, and Extended Sea and Inland Range.

General Power's views seemed not without justification when Admiral

Burke said in January 1959 that the Navy with OSD concurrence had projected a 15-year expansion plan for the PMR complex at an estimated total cost of $4 billion. The plan called for six separate firing ranges--Sea Test, IRBM, ICBM, Polar Orbit, Equatorial Orbit, and Anti-Missile-Missile.[30] The OSD-Navy plan was practically an unqualified adoption of Aerojet's recommendation.

To USAF officers at Vandenberg and at PMR, it appeared that the Navy was consistently trying to establish control over Vandenberg. This was brought out in numerous proposals that followed in rather rapid sequence-- requests for the physical transfer of Vandenberg missile assembly buildings to PMR, the move to acquire temporarily the use of 400 acres from Vandenberg's area, the use of Vandenberg airstrips by PMR aircraft. The contentions went on into the spring.[31]

With each week it became more and more evident that the Burke-White paper of March 1958 was no longer applicable to a rapidly changing situation. The Navy then prepared a proposed triservice-NASA agreement. The Air Force rejected it because its provisions threatened control of Vandenberg operations that "impugned directly on the future of the Strategic Air Command in space." Negotiations continued, however, at the chief of staff level. Before a satisfactory arrangement was reached, the dispute at PMR and Vandenberg became yet more unpleasant in connection with the "Discoverer crisis" in August 1959. Discoverers I and II were successfully launched on 28 February and 13 April 1959, but Discoverer III and IV, launched 3 and 25 June, both failed to orbit. The Air Force wanted to move the exit azimuth for future Discoverer launchings eastward to take advantage of the earth's rotation. PMR's commander claimed that the shift would endanger civilian life and property; the Air Force replied that the chance was 1 to

1,000,000. The question still remained unsettled when Admiral Burke and General White signed the new Navy-USAF "Agreement for Coordinated Peacetime Operation of the Pacific Missile Range" on 22 September. The agreement represented a compromise in the Navy's favor,* but the understanding had already been "overtaken by events" that were favorable to the Air Force.[32]

A Time for Decision

During the spring and early summer of 1959, interservice tension mounted. The criticism of the doctrine of aerospace, the PMR dispute, and the expressed desire of the Air Force to regain managerial control of research and development for its space projects were indicative of the prevalent restlessness. Moreover, it was time for changes in priority of projects. Some of those accorded highest national priority for research and development in January 1958--Thor, Jupiter, Jupiter-C, and Vanguard--could be removed from the list in March 1959 as having passed beyond that stage. On the other hand, Midas, Discoverer, and Sentry had advanced to the place where they should be accorded highest priority status. Soon the Secretary of Defense would have to make assignments of operational control for these satellites. Clearly a time for decision was at hand.[33]

---

*The agreement listed three major responsibilities of the Navy--range safety criteria, including approval of safety plans, procedures, and equipment for all missiles, satellites, and space vehicles launched at PMR; coordination to prevent duplication of range facilities and equipment; and the provision, maintenance, and operation of all common-use facilities, including ground instrumentation and the equipment required by joint tenancy agreements. Air Force rights were protected by the agreement to reserve for the service sponsoring the flight the control of flight preparation, the launching devices, and the missiles, satellites, and space vehicles themselves while in flight until the impact of missile or until the last-stage burnout of satellites and space vehicles.

An Air Force Attempt to Force the Issue

Although the Chief of Staff decided in the late winter of 1958 not to request a redefinition of service roles and missions, the decision did not preclude the request for operational assignments as satellite projects approached the end of development. Since it had been assumed that the Sentry reconnaissance satellite would eventually be placed under USAF operational control, the Under Secretary of the Air Force informally asked the Secretary of Defense on 26 February 1959 to approve the transfer at once. The Secretary of Defense agreed to consider the transfer if an official recommendation were made. By 15 April 1959 the Under Secretary of the Air Force had a recommendation from the Air Staff that the transfer of Sentry from ARPA to USAF be effected 1 July.[34]

The request for operational control of Sentry was extended to cover Midas too, and Headquarters USAF was optimistic. The field agencies, ARDC and AFBMD, apparently were not aware of what was happening, and on 18 May, General Schriever, ARDC's commander, wrote identical letters to Generals Gerhart and Wilson containing the following paragraph:[35]

> It is important that we get all or part of the space mission assigned to the Air Force as soon as possible including operational as well as development aspects. The Air Force is expending a great deal of time on space efforts much of which could be made more productive if the military space mission were clearly assigned. We could then pursue both our development and operational space efforts more aggressively. Moreover, such an assignment would do much to reduce the extensive and detailed "assistance" we have been getting from the Department of Defense, and place us in a better position relative to the National Aeronautics and Space Administration. It would permit us to plan and integrate our resources more effectively. It would also do much to clear the air between the Services and reduce the reactive efforts that take up so much of our time and keep us constantly off balance.

The two deputy chiefs of staff, in their reply to General Schriever, did little more than hint at the situation developing in the Pentagon.

Two weeks later, when Schriever attended a briefing there, he learned of the effort the Air Force had made to recover some of its lost projects. He learned also of a new position paper, completed 4 June, derived from the doctrine of aerospace. It stated that the Air Force would seek management responsibility for the research and development projects vital to space dominancy, request recognition as the executive agent of the Department of Defense for coordination and integration of research and development facilities and resources, and ask assignment of space systems to SAC and NORAD when operational.[36]

The Navy's Appeal for a Space Command

The Army and Navy were aware, of course, of the Air Force move to obtain assignment from the Secretary of Defense of Sentry and Midas, and there was every reason to believe that, in view of McElroy's statement in the Armed Forces Policy Council meeting, on 26 February, he would reach a decision in the near future regarding the disposition of the two projects.

On 22 April 1959, Admiral Burke suggested to the Joint Chiefs of Staff that the indivisibility of space and "the prospective magnitude of astronautical operations" required the establishment of a general military space agency. It would be under the direction of JCS and responsive "to the operational requirements of ARPA and NASA." Burke argued that the national interests would be served if "all facilities and functions" applicable to space vehicles and satellite operations, "including those of the three national missile ranges," were coordinated under a single command.[37]

The proposal ran counter to Air Force thinking that satellites should be operated by the service of primary interest, and it struck sharply against the doctrine of aerospace. The Chief of Naval Operations knew that his recommendations would be opposed by the Air Force, but he also knew that they would be approved by the Army and that he could count on the old Army-Navy advantage of 2 to 1 in JCS recommendations. Burke was therefore scarcely surprised when Gen. Maxwell Taylor, Chief of Staff, USA, gave full concurrence. The admiral must also have expected General White's objection to such a command taking over the "functions" of military space operations. White argued that the responsibility should go to the unified and specified commands. The Joint Chiefs referred the question to the Joint Staff with a request for recommendations by 15 June.[38]

ARPA's Move for a Mercury Joint Task Force

On 25 May, while the Joint Staff still considered the Navy's proposal, Dr. Glennan requested DOD assistance in the tracking and communication program for Mercury man-in-space flights.[39] The Secretary of Defense handed the question to a six-member ARPA-DDR&E-NASA committee that promptly turned to the Navy's proposal for a joint military space command and the objections of the Air Force to such an organization. There was thus a direct if unintentional connection between the Navy's proposal and ARPA's suggestion, after the joint committee completed its deliberations, for a Mercury joint task force. The Air Force concurred with ARPA but assumed that the task force would remain under ARPA, not JCS. The Air Force recommended that, since the needed facilities were part of the Atlantic Missile Range, the AMR commander should also serve as task force commander.[40]

On 30 June the Department of Defense informed Glennan that the military would support Mercury with a task force and appointed Johnson, in cooperation with Glennan to work out an initial plan. In a meeting of the ARPA, Army, Navy, and Air Force ad hoc group, all but the USAF representatives advocated a task force under JCS that could also serve as the nucleus of the space command proposed by the Navy. Despite Air Force protests, the ad hoc group requested ARPA to prepare a paper with this recommendation for service concurrence and presentation to the Armed Forces Policy Council for approval. The proponents of the task force held their point and went so far as to draft and redraft a DOD directive to that effect even after NASA signified a preference for a much simpler "coordination group" under AMR's commander.[41]

Suddenly the situation changed. As a result of NASA's recommendation and some personal intervention at the OSD-OSAF level, on 24 July the ARPA ad hoc group released a draft directive much nearer USAF desires. It designated Lt. Gen. Donald N. Yates, the AMR commander, as DOD representative for Mercury support and made him responsible to the Secretary of Defense. His additional duties included the preparation of plans to support the operation; direction and control of assigned DOD facilities, forces, and assets; and the furtherance of DOD specific missions to aid the project.[42]

Meanwhile, the Secretary of Defense had shown something less than enthusiasm for a joint space command.

McElroy's Decisions of 18 September 1959

In the midst of discussions concerning USAF's transfer request, Burke's joint-command proposal and NASA's Mercury support requirements, McElroy on 29 May 1959 asked JCS to recommend assignment of operational responsibility for Sentry, Midas, Transit, and the interim satellite detection system. McElroy asked that a reply be "expedited," but on 24 July the Joint Chiefs informed him that they could not reach agreement, and the question then came to lodge on his desk. After another three weeks, McElroy took a new approach. He personally briefed the Joint Chiefs on 13 August, to inform them of his "thinking." He deemed the scheduled satellite launchings up through 1963 too few to justify a joint military astronautics command "at the present time." He was also inclined to make the Air Force responsible for all DOD boosters, with payloads divided among the three services. He asked the Joint Chiefs for their "comments."[43]

The result was the same--the Joint Chiefs could not agree. McElroy then made the decision on his own. On 18 September the Secretary of Defense informed JCS that there would be no joint military space command; that the Air Force was responsible for launching DOD's space boosters; that the Air Force would have management responsibility for Sentry (Samos) and Midas; and that the Army and Navy would have similar responsibility for Courier and Transit respectively.[44]

Two years after Sputnik the tide seemed to have turned in favor of the Air Force, though the Secretary of Defense had not yet given the services managerial control of their research and development projects. There was reason neither for optimism or depression, for none could say what would be next. In the world of defense, as in the world of everything else, the "last" decision is never final.

*　　*　　*　　*　　*

The progress made by the American space program during the first two years after Sputnik could have been considered remarkable under different circumstances. There was, however, a disturbing pattern in overall events. For every cluster of small American accomplishments there was, without fuss or furor, a surpassing Russian achievement. There was unpublicized discontent within the National Security Council as evidenced by its policy statement of January 1960. There was criticism, too, on Capitol Hill, and among columnists and journalists. An example was the editorial that appeared in <u>Time</u> on 19 January 1959, just 17 days after the Russians launched their Lunik I on an interplanetary probe that sent "the first man-made planet circling the Sun with an estimated life of millions of years." Continuing, <u>Time</u> commented:

> Just when U.S. space achievements were beginning to make up for Sputnik jolts to the U.S. pride and prestige, the Russians sent their Lunik soaring far beyond where any man-made object had ever penetrated before. Once again the world marveled at U.S.S.R.'s technological prowess. Pressing an immediate question: Why is the U.S. still lagging in a race that may decide whether freedom has any future?

The answer to the question was not simple.

The basic element in the lag was the long period after World War II when civilian authorities failed to comprehend that the life of the nation could depend on an endlessly progressive technology. In the view of the House Select Committee on Astronautics and Space Exploration:[45]

> In the space field, in fact, the military people have generally shown far more foresight than the civilians, far more concern for applied science although still slow to appreciate the values of basic research. If the United States military mind was slow to grasp the worth of scientific discovery in the years leading into World War II, just the opposite has been true of the postwar decade. It is no accident, for example, that 50 percent of the physics doctorates in the United States since World War II were at least partially subsidized by the Office of Naval Research. Similar accomplishments could be quoted for the Army and Air Force. In fact, the military often had to hide valuable research work done under its aegis from the vengeful eyes of civilian budget experts. If the sputniks caught the United States by surprise, it was not for lack of warning from our military scientists.

It was this scientific conservatism at policy-making levels of the Government that gave the Soviets their head start in developing high-thrust rocket engines to serve with equal ease either an ICBM or a space program. Once handicapped by the Soviet time lead, there were technological factors that made American recovery seem slow indeed. Each new engine, for instance, required countless manhours for its design, development, construction, and testing. There was no easy ratio between chamber dimensions and thrust. General Wilson, DCS/Development, phrased it nicely when he said: "Propulsion is the key to space use. Up to the present we have not learned how to scale up a missile propulsion system to increase its thrust. Thus, each program must be undertaken as a separate and distinct development effort."[46]

Among the nontechnical factors of delay, none was more important than the lack of a stimulating and unifying national objective. Other than a space-for-peace policy, the cry was seldom for anything more than the dull motif of "catching up with Russia." An exception was voiced by Representative James G. Fulton, of Pennsylvania, in discussions with Livingston T. Merchant, Under Secretary of State for Political Affairs. Representative Fulton:[47]

> /I want to ask about the space program/ the administration is entering into. Are they simply trying to catch up to Russia in some fields or are they trying to keep ahead in others, or are we really going to have a program that I am for, of leapfrogging Russia? Would it be possible for us to have a space program that leapfrogged Russia and moved ahead our own targets more or less independently of her propaganda? Why don't we do that? Why don't we set targets ahead 3 to 5 years, far-reaching and far-seeking constructive targets, and then go ahead and reach them instead of looking to see how Russia is running and then run down that street?

To these penetrating questions the representative of the State Department replied: "I think it is a very constructive approach, sir."

The space-for-peace policy itself, though widely supported as an ideal, was sometimes criticized sharply because it divided the American space program into two unsynchronized parts--one that sought to move with the tempo of military necessity and one that would progress with the philosophic calm of pure science. Again to quote *Time*, many of the Administration's "scientific brains . . . proved to be nay-sayers and quibblers, among other things stirring up a futile, irrelevant dispute over whether space is a civilian or a military realm." *Time* regretted that the President's high-minded policy of space-for-peace did not stand up before the argument that great military advantages would be won by the nation first to make space its own "backyard."

These advantages seemed still irrefutable at the end of 1959, even though the military could not yet envision space tactics and strategy as clearly as Brig. Gen. William L. Mitchell foresaw the tactics and strategy of airpower when he bombed and sank the *Ostfriesland* off the Virginia shores in 1921. True, Mitchell had behind him experiences of World War I. The weakness of the military in 1959 lay in their inability to speak in more definite terms than the "uses of space" and categories of offensive and defensive space weapons. Yet nothing could refute the argument that space, by its very immensity, was certain eventually to introduce new concepts of warfare and weaponry.

He who comes second best in the space race will have no second chance to win.

NOTES

CHAPTER I

1. Presentation to Air Council by W.E. Dornberger, 7 Dec 57, History of Peenemunde.

2. W.D. Dornberger, V-2, James Cleugh & Geoffrey Halliday, trans (New York, 1954), pp 16-17.

3. Dwight D. Eisenhower, Crusade in Europe (New York, 1948), pp 258-60.

4. Dornberger, Hist of Peenemunde.

5. Statement by W. von Braun, ABMA, in House Hearings before the Select Cmte on Astronautics & Space Exploration, 85th Cong, 2d Sess, (Hereinafter cited as McCormack Cmte), Astronautics and Space Exploration, pp 19-20.

6. The United States Air Force Dictionary (Air University Press, 1956).

7. Von Braun before McCormack Cmte, p 17.

8. Brochure, SAC Space Concepts, 13 Aug 59; Fred Whipple, "Interview," Missiles and Rockets, Mar 57.

9. Press Conf with Charles E. Wilson, SOD, NY Times, 13 May 53.

10. Press Conf with Wilson, NY Times, 7 Jun 55.

11. J.H. Douglas, NY Times, 28 Apr 53; address by Sen Stuart Symington, Congressional Record, 23 Jun 53; Gen T.D. White, VC/S USAF, NY Times, 21 May 55; L/Gen D. L. Putt, DCS/D, USAF, NY Times, 14 Jun 55; NIE, 6 & 12 Nov 55.

12. NSC Action 915, 24 Sep 53, & 1433, 8 Sep 55.

13. Trevor Gardner, NY Times, 12 Aug & 9 Dec 55.

14. Wilson, NY Times, 7 Oct & 23 Nov 53.

15. NY Times, 9 Feb 56.

16. F.J. Malina, "Is the Sky the Limit?" Army Ordnance, Jul-Aug 46; H.E. Newell, Jr, High Altitude Rocket Research (New York, 1953), p 13.

17. The Development of the Navaho Missile by J.A. Neal, WADC Hist Br, Jan 55, p 1; ltr, M/Gen B.W. Chidlaw, Dep CG/Engineering, AMC to CG

AAF, 6 May 47, subj: AAF GM Program, Doc 73 in AMC History of the Development of Guided Missiles, 1946-1950, by M.R. Self.

18. Self, pp 42-43.

19. Ltr by Chidlaw, 6 May 47.

20. Memo, B/Gen T.S. Power, Dep AC/AS-3 to CG AAF, 16 Jun 57, subj: Operational Rqrmts (Priorities) for GM's, 1947-1957, in Self, Doc 76.

21. Neal, pp 1-15.

22. Ibid., pp 4, 28-31.

23. C.H. Donnelly, The United States Guided Missile Program, prep for Preparedness Investigating Subcmte of Senate Cmte on Armed Services, 86th Cong, 1st Sess (Washington, 1959), pp 62-69.

24. Ltr, T. Gardner to Lewis Strauss, 14 Jul 54, ARDC GO 42, 15 Jul 54, subj: Reorgn of Hq ARDC & Orgn of WDD.

25. Rpt of Technological Capabilities Panel (TCP) of the Science Advisory Cmte, Meeting the Threat of Surprise Attack (Washington, 14 Feb 55), passim.

26. Donnelly, pp 81-82.

CHAPTER II

1. Dornberger, V-2, pp 16-17.

2. Ltr, Col L.A. Hall, D/Warning & Threat Assessments, AC/I to M/Gen G.W. Martin, D/Plans, 18 Dec 59, subj: Organization, Planning, & Control of the Soviet Space Prog; interview, Mae Link, Max Rosenberg, & Lee Bowen with B/Gen D.D. Flickinger, 25 May 61.

3. Staff Rpt, House Select Cmte on Astro & Space Exploration, Astronautics and its Applications (Washington, 1958) (Hereinafter cited as Staff Rpt), p 223.

4. NSC 5814/1, U.S. Policy on Outer Space, Anx A, The Soviet Space System, 18 Aug 58.

5. Memo, K.T. Compton, Chmn/RDB to SOD, 31 Oct 59, subj: Progress Rpt Study of GM Prog.

6. Wash Evening Star, 7 Sep 51; NY Times, 4 Oct 51; Wash Post, 12 Oct 51. The reports multiplied during the next 12 months.

7. NSC 5814/1, Anx A; Staff Rpt, pp 217-37; House Rpt 1758, Select Cmte on Astro & Space Exploration, 85th Cong, 2d Sess, The National Space

Notes to pages 28-36.

Program, App 5, *passim*; A.T. Zachringer, "Soviet Astronautics," *Missiles & Rockets*, Feb 57.

8. Staff Rpt, p 221.

9. NSC 5814/1; H Rpt 1758, p 31.

10. "General Arnold's Third Report," *The War Reports of General of the Army George C. Marshall, General of the Army H.H. Arnold, and Fleet Admiral Ernest J. King* (Philadelphia & New York, 1947), p 463.

11. Ltr, R.C. Lenz, Jr, D/Planning, DCS/P&O, WADD to Col C.D. Gasser, SAB, 7 Apr 60, AF Space Prog, Atch 4; Douglas Aircraft Co Contract W-33-038, ac-14105; B/Gen B.H. Schriever, Briefing at Pentagon on ARDC, 24 Jan 54.

12. E.M. Emme, *Aeronautics and Astronautics* (Washington, 1961), p 61.

13. Rpt of Technical Evaluation Gp Satellite Vehicle Program (GM 13/70), 29 Mar 48 (Hereinafter cited as TEG Rpt); Capt C.W. Styer, Jr, USN, & Comdr R.F. Freitag, USN, "The Navy in the Space Age," *U.S. Naval Institute Proceedings*, Mar 60.

14. Memo, Ch/BuAer to JRDB, 24 Jan 47, subj: Earth Satellite Vehicle, Incls A&B.

15. Rand RA-15000, *First Quarterly Report*, Jun 46, pp 6-17.

16. Ibid.; Douglas Rpt SM-11827, Preliminary Design of An Experimental World Circling Spaceship, 2 May 46.

17. Douglas Rpt SM-11827, pp 2, 9-16.

18. Memo, R/Adm L.C. Stevens, Chmn/R&D Cmte, Aeronautical Bd to Aero Bd, 15 May 46, subj: Case No. 244--High Altitude Earth Satellite; Aero Bd Minutes, 29 May 46.

19. Memo, Ch/BuAer to JRDB, 24 Jan 46; JRDB Rules of Orgn and Procedure, 7 Oct 46.

20. JRDB Minutes of 6th Mtg, 6 Mar 47, w/atch AAF Position Papers; memo, L/Gen H.S. Vandenberg, Senior Army Member of Aero Bd to JRDB, 13 Jun 47, subj: Satellite Agency.

21. JRDB Minutes of 6th Mtg; memo, RDB to Aero Bd, 9 Jan 48, subj: Earth Satellite Vehicle.

22. Vannevar Bush, *Modern Arms and Free Men* (New York, 1949), pp 84-86.

23. CGM Minutes, 3 Feb 48, Item 10; ltr, F.L. Hovde, Chmn/CGM to W.A. McNair, Chmn/TEG, 6 Feb 48.

24. TEG Rpt.

25. Ltr, M/Gen H.B. Saylor, Ch/R&D Div, Ord Dept to Secy/CGM, 26 Jul 48, subj: Comments on TEG Rpt; ltr, R/Adm D.V. Gallery, ACNO (GM) to Secy/CGM, 8 Jul 48; ltr, M/Gen F.O. Carroll, USAF Member/CGM to Chmn/CGM, 27 Jul 48, subj: Satellite Vehicle Prog; CGM Minutes, 15 Sep 48, Item 12.

26. First Report of the Secretary of Defense, 1948, App C, Rpt of the Exec Secy/RDB, p 129.

27. W.B. Bergen, Pres/Martin Co, quoted in Denver Post, 25 Sep 59.

28. H.A. Zahl, Research D/Army Signal Corps, quoted in Wash Evening Star, 28 Oct 58; account of incident in NY Tribune, 18 Oct 60.

29. Ltr, B/Gen A.R. Crawford, Ch/Engr Div, AMC to DCS/M, Hq USAF, 8 Dec 47, subj: Project Rand Satellite Vehicle; memo, L/Gen H.A. Craig, DCS/M to VC/S, 12 Jan 48, subj: Earth Satellite Vehicle.

30. Rand R-217, Utility of a Satellite Vehicle for Reconnaissance (Project Feed Back), Apr 51.

31. Memo, Col J.A. Dunning, Off of D/Plans to D/Plans, 28 Oct 57, subj: USAF Satellite Prog.

32. Rand Rm-1194, Scientific Uses for a Satellite Vehicle, 12 Feb 54; Rand Rm-1500, Scientific Use of an Artificial Satellite, 8 Jun 55.

33. Hist, AFFTC, Jan-Jun 58, pp 164-82.

34. Rand R-217.

35. R&R Project Card, ARDC to Hq USAF, 10 Oct 57, subj: Abbreviated System Plan.

36. Memo, Col D.L. Kime, Off of D/Dev Plang to D/R&D, 18 Nov 57, subj: Comments on Abbr System Dev Plan; M/R by Col B.H. Ferrer, 4 Nov 57, subj: Boost Glide Concept; D/R&D Dev Directive (DD) 94, 25 Nov 57, subj: Hypersonic Strategic Weapon System.

37. RCA-Rand RM-999, Progress Report (Project Feed Back), 1 Jan 53.

38. Rand R-262, Project Feed Back Summary Report, Vols I & II, 1 Mar 54.

39. Hq ARDC Briefing to SAB Ad Hoc Cmte on Advanced Weapons Tech & Environment, Space Technology, 29 Jul 57, pp 17-21.

40. Hq ARDC Hyper-Environment Test System Dev Plan, 18 May 59, p 1-1.

41. Ibid.

42. Memo, D/R&D to R.E. Horner, ASAF (R&D), 13 May 57, subj: BALWARDS Vehicles; memo, Hq USAF to Hq ARDC, 15 May 57, subj: BALWARDS.

Notes to pages 50-62.

43. Memo, Hq USAF to Hq ARDC, 3 Feb 58, subj: BRATS; Hq ARDC Hyper-Environment Test System Supporting Plan, 13 Mar 59.

44. J.S. Hanrahan & David Bushnell, Space and Biology (New York, 1960), pp 3-14.

45. B/Gen D.D. Flickinger, D/Human Factors, Hq ARDC, "Human Factor Problems in Space Flight," in Space Technology, an ARDC presentation to SAB, 29 Jul 57, pp 57-67.

46. AFMDC, History of Research in Space and Biodynamics, 1946-1958, pp 1-28.

47. Off of Asst/Bioastronautics, Hq ARDC, Capability & Rqrmts for Manned Space Ops, 15 Dec 60.

48. Flickinger; memo, Gen T.S. Power, CINCSAC to C/S USAF, 13 Aug 58, subj: SAC Space Policy.

49. Hist, AFMDC, Jul-Dec 57, pp 29-52.

50. Memo, Col J.L. Martin, Dep D/AT to Off/LL, 10 Nov 59, subj: Project Mercury, w/atch 1, Outline of History of USAF Man-in-Space; Hq ARDC, Bioastronautics Capability & Rqrmts for Manned Space Ops, 15 Dec 60, pp 4-5.

51. Ezra Kotcher, Off of D/Labs, WADC, "Vehicle Design," in Space Technology (presentation), pp 25-36.

52. Statement by R/Adm J.T. Hayward, ACNO (R&D) before McCormack Cmte, p 275.

CHAPTER III

1. H Rpt 1758, 85th Cong, 2d Sess, p 19.

2. J.C. Cooper, "High Altitude Flight and National Sovereignty," International Law Quarterly Review, Spring 58.

3. Sir A.D. McNair, The Law of the Air (London, 1953), App, pp 295-328; Col M.B. Schofield, USAF, "Control of Outer Space," Air University Quarterly Review, Spring 58.

4. Space Law, a symposium prep at request of Sen L.B. Johnson, Chmn/Senate Sp Cmte on Space & Astronautics, 85th Cong, 2d Sess, 31 Dec 58, passim; Sir Geoffrey Butler & Simon MacColey, The Development of International Law (London, 1958), pp 40-58.

5. Oscar Schacter, "Who Owns the Universe?" in Space Law, pp 8-17.

6. Wenner Buedeler, The International Geophysical Year (UNESCO, 1957), pp 14-26, 48-49.

7. Joseph Kaplan, The IGY Rocket and Satellite Program (Washington, 1960), p 13.

8. NY Times, 16 Apr 55.

9. W.B. Hester, "Some Political Implications of Space Flight," Journal of the British Interplanetary Society, Nov-Dec 55.

10. TCP Rpt Meeting the Threat, pp 146-47.

11. NSC 5520, 26 May 55.

12. R.W. Young, "The Aerial Inspection Plan and Air Sovereignty," George Washington Law Review, Apr 56.

13. Maj H.J. Neumann, USAF, "The Legal Status of Outer Space and the Soviet Union," in Space Law, pp 495-503. Neumann cited A. Kislov and S. Krylov, "State Sovereignty in Air Space," International Affairs (Moscow), Mar 56.

14. NY Times, 30 Jul 55.

15. Facts on File, 28 Jul-3 Aug 55, p 255.

16. NY Times, 8 Feb 56.

17. J.C. Cooper, "Legal Problems of Upper Space," Proceedings of American Society of International Law, 50th Annual Meeting, 25-28 Aug 56; NY Times, 11 Jan 57.

18. NY Times, 15 Jan 57; Yearbook of the United Nations, 1956 (New York, 1957), pp 100-101.

19. H Doc 372, 84th Cong, 2d Sess, Rpt of President to Cong for 1957, U.S. Participation in the United Nations, pp 14-15.

CHAPTER IV

1. Press Conf with Wilson, NY Times, 21 Nov & 17 Dec 54.

2. TCP Rpt Meeting the Threat, pp 146-47.

3. Draft memo, SAF to SOD, 11 Apr 55, subj: TCP Rpt.

4. Memo, C.E. Wilson to Secys Army, Navy, & AF, 28 Mar 55, subj: Earth Satellites.

5. Memo, JCS to SOD, 18 Apr 55, subj: TCP Rpt, JCS 1899/200.

6. NSC 5520, 26 May 55.

Notes to pages 72-78.

7. Memo, D.A. Quarles, ASOD (R&D) to Coordinating Cmte on Gen Sciences, 30 Mar 55, subj: Scientific Satellite; M/R, A.E. Lombard, Scientific Adviser, D/R&D, 5 Apr 55, subj: Scientific Satellite.

8. Memo, R.W. Cairns, Chmn/Coord Cmte on Gen Sciences to ASOD (R&D), 4 May 55, subj: Scientific Satellite Prog in the DOD; memo, D.A. Quarles to SAF, 17 May 55, subj: Scientific Satellite; Rpt of Ad Hoc Advisory Gp on Sp Capabilities, 4 Aug 55, App F.

9. Memo, Col Joseph W. Williams, Dep Asst/Dev Prog to M/Gen H.C. Donnelly, Sp Asst to C/S, 18 Mar 58, subj: Project World Series; Rpt of Ad Hoc Advsy Gp.

10. Rpt of Ad Hoc Advsy Gp.

11. Ibid.

12. Paper prep by DOD for House Select Cmte on Astro & Space Exploration, 85th Cong, 2d Sess (McCormack Cmte), pp 155-56; L/Gen J.M. Gavin (Rtd), War and Peace in the Space Age (New York, 1958), pp 14-15.

13. Memo, R.B. Robertson, Jr, Actg Dep SOD to Secys/Army, Navy, AF, 9 Sep 55, subj: Tech Prog for NSC 5520; memo, J.B. McCandery, Dep ASOD (R&D) to ASAF (R&D), 19 Sep 55, subj: AF Scientific Satellites; Statement by R/Adm J.T. Hayward, ACNO (R&D) before McCormack Cmte.

14. "Project Vanguard," Missiles and Rockets, Jul 57.

15. Statement by W. von Braun before McCormack Cmte, p 63.

16. C.C. Furnas, ASOD (R&D), quoted in NY Times, 16 Oct 57.

17. C.C. Furnas, quoted in Facts on File, 10-16 Oct 57, p 331.

18. Rand RM-1922, A Casebook on Soviet Astronautics, Pt II, 21 Jun 57, passim; ASTIA Doc AD 133038, "Soviet Space Flight Activities," a discussion of the Rand space flight program, presented to AFSAB by F.J. Krieger, 29 Jul 59.

19. Memo, P.F. Bundage, D/BOB to the President, 30 Apr 57, subj: Project Vanguard.

20. Ibid.; memo, J.S. Lay, Jr, Exec Secy/NSC to members of NSC, 3 May 57, subj: U.S. Scientific Satellite Prog; memo, C.E. Wilson to JCS & Secys/Army, Navy, AF, 17 May 57, subj: U.S. Scientific Satellite Prog.

21. Von Braun before McCormack Cmte, pp 33-34, 66.

22. Statement by Gen Gavin before McCormack Cmte.

23. Von Braun before McCormack Cmte, pp 18-19, 34.

24. Gavin before McCormack Cmte, pp 184-85.

25. "Project Vanguard," *Missiles and Rockets*, Jul 57.

26. S.F. Singer, "Synoptic Rocket Observations in the Upper Atmosphere," *Nature*, 20 Jun 53.

27. Memo, Col W.O. Davis, AFOSR to Col D.D. Flickinger, 30 Jul 54, subj: Program Mouse.

28. Memo, Col Davis to F.M. Field, D/R&D, 12 Jul 57, subj: Far Side Project Justification.

29. Memo, Col B.H. Holzman, ARDC to M/Gen M.C. Demler, 20 Mar 57, subj: Project Far Side.

30. Ltr, B/Gen A.D. Starbird, AEC to AFOSR, 17 Apr 57; ltr, U-Secy, Dept/Interior to SAF, 21 Jun 57.

31. Aeronutronics System Pub C-100, 16 Sep 57, subj: Conduct of Far Side Phase II Expers.

CHAPTER V

1. H Rpt 1371, 90th Rpt by Cmte on Govt Ops, 85th Cong, 2d Sess, *U.S. Military Aid and Supply Programs in Western Europe*, p 31.

2. Ibid.; H Rpt 1121, 11th Rpt by Cmte on Govt Ops, 86th Cong, 1st Sess., *Organization and Management of Missile Programs*, pp 1-2.

3. *Astronautics*, Nov 57. This comment was typical and could be supported by references to numerous journals and periodicals.

4. *Facts on File*, 10-16 Oct 57.

5. These and other comments may be found in *NY Times*, *Wash Post*, *Facts on File*, etc., for period 5 Oct-7 Dec 57.

6. G. Zadoroghnyi, "The Artificial Satellite and International Law," *Sovetskaia Rossiia*, 17 Oct 57, trans by A.M. Jones, in *Space Law*, pp 504-6.

7. Von Braun before McCormack Cmte.

8. Statement by R/Adm Hyman G. Rickover before McCormack Cmte, p 261.

9. Statement by Loftus Becker, Legal Adviser, Dept/State, before McCormack Cmte, pp 1269-75.

10. Dept/State *Bulletin*, 9 Jun 58.

11. H Rpt 1769, 85th Cong, Exploration of Outer Space, passim; S Rpt 1728, 85th Cong, Peaceful Exploration of Space, passim.

12. NY Times, 11 Dec 57, 11 Jan & 3 Feb 58.

13. Yearbook of the United Nations, 1958 (New York, 1959), pp 19-22.

14. Ibid.

15. H Doc 104, 85th Cong, 2d Sess, Rpt of President to Cong for 1958, U.S. Participation in the United Nations.

16. Ibid.

17. NY Times, 18 Nov 58.

18. H Doc 104, 85th Cong; Dept/State study, "Major Elements of United States International Consultative and Cooperative Activities in the Field of Outer Space," in H Hearings before Cmte on Science & Astronautics, 86th Cong, 2d Sess, Review of the Space Program, pp 28-32; Yearbook of the United Nations, 1959 (New York, 60), pp 24-28.

19. Dept/State, "Major Elements . . . of Activities in Outer Space."

20. NY Times, 8 Nov 57.

21. Ibid., 6 Feb 58.

22. Ibid., 13 Jan & 3 Apr 58; memo, JCS to SOD, 25 Nov 57, subj: DOD Sp Projects Agency; Rpt of President's Science Adv Cmte, Introduction to Outer Space, 26 Mar 58.

23. Memo, M/Gen J.B. Cary, D/Plans to C/S USAF, 26 Jun 58, subj: U.S. Policy on Outer Space; memo, Col R.F. Worden, Off of D/Plans to Gen Cary, D/Plans, 19 Jun 58, subj: National Policy on Space; NSC 5814/1, Preliminary U.S. Policy on Space, 18 Aug 58.

24. NSC 5814/1.

25. OCB Operation Plan for Outer Space, 18 Mar 59.

26. Memo, Col R.R. Rowland, Off of D/Plans to D/AT et al., 27 Aug 59, subj: U.S. Policy on Outer Space, NSC 5814/1.

27. NSC 5901/1, A National Security Policy, 5 Aug 57.

28. Memo, L/Col E.N. Kohrman, Off of D/Plans to C/S USAF, 18 Nov 59, subj: Final Draft of U.S. Policy on Outer Space.

CHAPTER VI

1. <u>Newsweek</u>, 18 Nov 57.

2. Trevor Gardner, "But We Are Still Lagging," <u>Life</u>, 4 Nov 57.

3. Statement by Edwin L. Weisl, Ch Counsel/DOD, in S Hearings before Preparedness Investigating Subcmte of Cmte on Armed Services, 85th Cong, 1st and 2d Sess (hereinafter cited as Johnson Subcmte), <u>Inquiry into Satellite and Missile Programs</u>, p 220.

4. Rpt of Teller Ad Hoc Cmte, Hq ARDC, 28 Oct 57.

5. DOD Dir 5105.10, 15 Nov 57; Statements of Neil McElroy & W.M. Holaday before Johnson Subcmte, pp 211-14, 346-49; H Rpt 1121, as cited in V, n 2, p 17.

6. Statement by Neil McElroy, in H Hearings before DOD Subcmte on Appropriations, 85th Cong, 1st Sess (hereinafter cited as Mahon Subcmte), <u>The Ballistic Missile Program</u>, p 7.

7. Memo, JCS to SOD, 25 Nov 57, subj: DOD Sp Projects Agency.

8. Memo, SOD to JCS, 6 Dec 57, subj: DOD Sp Projects Agency; H Doc 298, 85th Cong, 2d Sess, Comm from the President, "Proposed Additional Authority for the Department of Defense," 7 Jan 58; McElroy before Mahon Subcmte, p 7.

9. H Rpt 1121, pp 132-35.

10. DOD Dir 5105.15, 7 Feb 58, subj: ARPA.

11. Memo, P.D. Foote, W.M. Holaday, & R.W. Johnson to SOD, 7 Apr 58.

12. <u>NY Times</u>, 10 Jan 58.

13. Ibid., 4 Apr 58.

14. PL 85-599, DOD Reorganization Act of 1958, appr 6 Aug 58; memo, M/Gen H.C. Donnelly, Asst DCS/P&P to DCS/P&P, 4 Aug 58, subj: DOD Reorgn, w/2 Incls; <u>Facts on File</u>, 18-24 Dec 58.

15. Memo, Col V.Y. Adduci, Off/LL to VC/S, 22 Nov 57; memo, L/Gen D.L. Putt, DCS/D to M/Gen J.E. Smart, Asst VC/S, 10 Dec 57, subj: Estbmt of a Dir of Astronautics.

16. <u>NY Times</u>, 14 Dec 57; verbal info from Off of C/S thru Col J.L. Martin, D/AT to Lee Bowen, 27 Jul 60; memo, Putt to All Dirs <u>et al</u>., 13 Dec 57, subj: Cancellation of the Memo of 10 Dec 57.

17. Memo, Asst VC/S to Air Staff, 4 Mar 58, subj: Space Projects Involving ICBM/IRBM's.

Notes to pages 111-118.

18. Memo, M/Gen J.S. Mills, Asst DCS/D to Gen Smart, 12 Mar 58, subj: A Dir of Advanced Weapons; memo, B/Gen R.H. Warren, Mil Asst/SOD to SAF, 22 Jul 58, subj: Dir of Advanced Technology; DAF GO 44, 29 Jul 58, subj: Estbmt of AFDAT.

19. Memo, L/Gen R.C. Wilson, DCS/D to All Dirs et al., 29 Jul 58, subj: Estbmt of AFDAT.

20. Memo, B/Gen H.A. Boushey, D/AT to C/S USAF, 8 Dec 58, subj: Functions of AFDAT; memo, M/Gen C.M. McCorkle, Asst C/S (GM) to Asst VC/S, 13 Jan 59, subj: Function of AFDAT; memo, Smart to Deps et al., 6 Apr 59, subj: Respons for Space Projects; HOI 21-14, 13 Apr 59.

CHAPTER VII

1. Memo, Eilene Galloway to Rep J.W. McCormack, 7 Mar 59, subj: Problems of Cong in Formulating Outer Space Leg, in H Hearings before McCormack Cmte.

2. Ibid.

3. NY Times, 3 Apr 58.

4. H Rpt 1758, 85th Cong, 2d Sess, The National Space Program.

5. Statements before McCormack Cmte, by H.L. Dryden, D/NACA, p 403; C.C. Furnas, Chancellor, U. of Buffalo, pp 735-73; Lee DuBridge, Pres, Calif Institute of Tech, pp 778-79; W.H. Pickering, D/JPL, p 804; J.A. Van Allen, States Univ, Iowa, pp 864-65; W.S. Dornberger, Tech Asst to Pres, Bell Aircraft Corp, pp 1085-94.

6. Statements before McCormack Cmte by M/Gen John B. Medaris, AOMC, p 167; R/Adm Hyman G. Rickover, AC/BuShips for Nuclear Propulsion, pp 222, 230-31; R/Adm John T. Hayward, ACNO (R&D), p 276; L/Gen Donald L. Putt, DCS/D, USAF, pp 99-134; M/Gen Bernard A. Schriever, COMAFBMD, pp 647-57; H.F. York, Ch Scientist/ARPA, p 47; J.D. Doolittle, Chmn/NACA, p 928; R/Adm W.F. Raborn, D/Sp Projects, USN, pp 885-909. See also ltr, D.A. Quarles, Dep SOD to M.H. Sterns, D/BOB, 1 Apr 58.

7. PL 85-568, The National Aeronautics and Space Act of 1958, appr 29 Jul 58.

8. H Rpt 1758.

9. PL 85-568.

10. Ibid.

11. Ibid.

12. H Rpt 2166, National Aeronautics and Space Act of 1958 Conf Rpt, 15 Jul 58, p 16.

13. NASA Release, 1 Oct 58.

14. Memo, McElroy to Secys/Army, Navy, AF, 12 Sep 58, subj: The Civilian-Military Liaison Cmte.

15. Ibid.; M/R by M/Gen M.C. Demler, D/R&D, 23 Jul 58; memo, Demler to Asst DCS/D, Sep 58, subj: DOD-NASA Orgn & Relationship; memo, Quarles to Secys/Army, Navy, AF, 31 Oct 58, subj: C-MLC.

16. Memo, JCS to SOD, 22 Oct 58, subj: C-MLC, JCS 2283/17.

17. Memo, Secy/Army to D/ARPA, nd; memo, Secy/Navy to D/ARPA, nd; ASSS by B/Gen H.A. Boushey, Dep D/R&D, 8 Oct 58; memo, M/Gen R.P. Swofford, Asst DCS/D to Adm Asst/OSAF, 8 Oct 58; ASSS, L/Gen R.C. Wilson, DCS/D to All Dirs et al., 14 Oct 58, all same subj: Terms of Ref for C-MLC.

18. Memo, J.H. Douglas, SAF to C/S USAF, 16 Oct 58, subj: Terms of Ref for C-MLC; memo, Swofford to SAF, 24 Oct 58, subj: Initial Mtg of DOD Nominees to C-MLC; JCS 2283/18, Terms of Reference of C-MLC.

19. M/R by Maj H.C. Howard, 23 Oct 58, subj: C-MLC.

20. Memo, Wilson to D/AT et al., 29 Oct 58, subj: USAF Adm C-MLC Channels.

CHAPTER VIII

1. Interview, Joseph Angell, Max Rosenberg, & Lee Bowen with L/Gen D.L. Yates, 7 Mar 61.

2. H Rpt 1121, 86th Cong, 1st Sess, Organization and Management of Missile Programs, p 131.

3. Ibid., pp 131-32.

4. Yates interview.

5. Statement by W.M. Brucker, Secy/Army, in H Hearings before Cmte on Science & Astronautics, 86th Cong, 1st Sess, Missile Development and Space Sciences, p 200.

6. Astronautics Briefing to Armed Forces Policy Council by M/Gen J.S. Mills, Asst DCS/D, 5 Nov 57; memo, M/Gen R.P. Swofford, Jr, D/R&D to R.E. Horner, ASAF (R&D), 22 Nov 57, subj: Proposed DOD Dir of Sp Projects; memo, Horner to SAF, Attn: Sp Asst/Armed Forces Policy Council, 12 Nov 57, subj: Outer Space Vehicles; memo, William Weitzen, Off of ASAF (R&D) to Horner, 18 Dec 57, subj: Astronautic Planning; memo, J.H. Douglas, SAF to Neil McElroy, SOD, 12 Nov 57, subj: Respons for Mil Satellites.

Notes to pages 128-139.

7. Memo, W.A. Holaday, Sp Asst/SOD (GM) to Chmn/Advs Gp on Sp Capabilities, 6 Sep 57, subj: Satellite Progs.

8. Memo, Exec Secy/Advsy Gp on Sp Cpblities to ASAF (R&D), 11 Oct 57, subj: WS-117L.

9. Memo, J.H. Stewart, Chmn/Adm Gp on Sp Cpblities to Holaday, 15 Jan 58, subj: Satellite Plans of the Mil Depts, w/incl, p 20; Col R.C. Richardson, Asst/Long Range Objectives, DCS/P&P to M/Gen H.C. Donnelly, Asst DCS/P&P, 14 May 58, subj: NSC Policy on Outer Space.

10. Memo, Stewart to Holaday, 15 Jan 58, incl, pp 20-21.

11. *Ibid.*, p 21.

12. *Ibid.*, pp 3-12.

13. Memo, Holaday to SAF, 7 Jan 58, subj: Satellite Program.

14. Memo, Horner to Holaday, 25 Jan 58, subj: Air Force Astro Dev Prog.

15. Verbal comments by William Weitzen, Off of ASAF (R&D) to Lee Bowen, 12 Aug 60; tele con Lee Bowen with B/Gen H.A. Boushey, Comdr/Arnold Engr Dev Center, 25 Aug 60.

16. Memo, Douglas to SOD, 1 Feb 58, subj: Recon Satellites, w/incl, draft memo, SOD to SAF, subj: Prog for Mil Recon Satellite (WS-117L).

17. Memo, Douglas to SOD, 14 Feb 58, subj: Thor & WS-117L Prog.

18. Memo, Horner to SOD, 21 Feb 58.

19. Memo, McElroy to SAF, 24 Feb 58, subj: AF WS-117L Prog.

20. Memos, R.W. Johnson, D/ARPA to Secys/Army, Navy, & AF, 27 Mar 58.

21. Statement by Neil McElroy in H Hearings before DOD Subcmte on Appn, 85th Cong, 1st Sess, The Ballistic Missile Program, p 7; DOD Dir 5105.15, 24 Mar 58, subj: ARPA; memo, the President to SOD, 7 Feb 58, subj: Approval of Adv Space Projects; memo, Johnson to Secy/Army, 27 Mar 58.

22. AO 2, Amnd 4, 29 Sep 58; AO 9, Amnd 5, 29 Sep 58, & Amnd 6, 11 Dec 58.

23. Memo, Johnson to SAF, 28 Feb 58, subj: Recon Satellites & Manned Space Explor.

24. Memo, B/Gen H.A. Boushey, Dep D/R&D to M/Gen J.S. Mills, Asst DCS/D, 18 Mar 58, subj: Suggested Remarks for Col Oder at Presentation to Roy W. Johnson, 1000, on 19 Mar 58; memo, Mills to VC/S USAF, 20 Mar 58, subj: Man in Space Prog; memos, M.A. McIntyre, OSAF to D/ARPA,

19 Mar 58, subjs: AF Man in Space Prog & WS-117L.

25. Memo, Johnson to SOD, 19 Mar 58, subj: Proposed ARPA Projects, w/3 incls, ARPA Projects 1, 2, & 3.

26. Memo, McElroy to the President, 19 Mar 58, subj: ARPA's Proposed Projects; memo, the President to SOD, 24 Mar 58; AO's 1, 2, & 3, 27 Mar 58.

27. Memo, D.A. Quarles, Dep SOD to D/ARPA, 4 Apr 58, subj: Argus.

28. Memo, Quarles to D/ARPA, 1 May 58, subj: Satellite Progs, Including the Vanguard Series.

29. Memo, L/Gen D.L. Putt, DCS/D to C/S USAF, 3 Apr 58, subj: Recon Satellite.

30. Memo, Boushey to M/Gen J.E. Smart, Asst VC/S USAF, 2 Apr 58, subj: USAF Astro Prog, w/incl, Proposed Memo, Horner to D/ARPA, same subj.

31. Memo, L/Col F.J. Dillon, Jr to Col Leon Booth, Sp Asst/DCS/D, 10 Sep 58, subj: Air Staff Orientation; AO's 10-14 & 17-19, 25 Jul-29 Aug 58.

32. Ltr, Gen C.E. LeMay, VC/S USAF to L/Gen S.E. Anderson, COMARDC, 9 May 58.

CHAPTER IX

1. Ltr, Gen T.S. Power, CINCSAC to C/S USAF, 13 Aug 58, subj: SAC Space Policy.

2. Statement by H.F. York, Ch Scientist/ARPA, in H Hearings before Subcmte of Cmte on Appn, 85th Cong, 2d Sess, Department of Defense Appropriations, pp 295-319.

3. Statement by H.L. Dryden, D/NACA, submitted to the McCormack Cmte, pp 949-50.

4. R.S. Cesaro & Robertson Youngquist, "Strategic Space--Key to National Survival," Statement at 5th Annual ARDC Science & Engineering Symposium, Hq ARDC, 22 Jul 58.

5. AO 28-59, 29 Sep 58.

6. AO 2-58, Amnd 5, 6 Oct 58; AO 12-59, Amnd 1, 6 Oct 58, AO 10-59, Amnd 7, 13 Apr 59; AO 17-59, Amnd 5, 13 Apr 59.

7. M/R by Maj Paul L. Chell, Analysis Div, Hq ARDC, 24 Apr 59, subj: ABMA; Exec Orders 10783 & 10793, 1 Oct & 3 Dec 58; Statement by T.K. Glennan, A/NASA, in S Hearings before NASA Auth Subcmte of Cmte on Aero & Space Science, 86th Cong, 1st Sess, (hereinafter cited as

Notes to pages 150-158.   219

Stennis Cmte), <u>NASA Authorization for Fiscal Year 1960</u>, Pt 1, Scientific and Technical Presentations, pp 6-7.

8. Rpt., NACA Sp Cmte on Space Tech, 28 Oct 58, subj: <u>Recommendations to NASA for a National Space Program</u>, passim.

9. Stennis Cmte, Pt 2, Program Detail for Fiscal Year 1960, passim.

10. Ltr, D.A. Quarles, Actg SOD to Glennan, 10 Dec 58; memo, Quarles to Secys/Army, Navy, & AF, 10 Dec 58, subj: Prelim Study of the Status of the Large Rocket Progs, w/Atch 1; NASA Rpt to the President, 27 Jan 59, subj: A National Space Vehicle Prog.

11. NASA Rpt to the President, 27 Jan 59.

12. Stennis Cmte, Pt 1, passim.

13. NASA Presentation to C-MLC, 15 May 59, subj: National Space Sciences Prog.

14. Statements by W.E. Moeckel & J.H. Childs before Stennis Cmte, Pt 1, pp 36-40, 48-54.

15. Glennan before Stennis Cmte, Pt 1, p 6.

16. Memo, Col J.L. Martin, D/AT, to M/Gen R.P. Swofford, 5 Jan 59, subj: Presentation for Comdrs Conf; ltr, H.F. York, DDR&E to C.F. Ducander, Ch Counsel/H Cmte on Science & Astro, 30 Dec 59.

17. Martin to Swofford, 5 Jan 59.

18. M/R by L/C K.G. Lundell, 4 Jun 58, subj: System 609A BMTS; ltr, Dryden to Boushey, 11 Jul 58; Memo of Understanding, 31 Oct 58, subj: Coop between NASA & the Air Force in the Dev of a Solid-Rocket Test Vehicle; ltr, L/Col J.R. Ryan, Astro Div, Hq ARDC to DCS/D, Hq USAF, 18 Dec 58, subj: System 609A.

19. Stennis Cmte, Pt 2, pp 743-48; memo, Col G.B. Knight, USAF to M/Gen W.P. Fisher, subj: Testimony by NASA Witnesses, 21 May 59.

20. Hist AMC's Ballistic Missile Center, Jan-Jun 59, I, pp I [sic], 48-52.

21. Ibid.; memo, Schriever to C/S USAF, 2 Feb 59, subj: NASA/AF Operating Procedures.

22. Hist, AMC's Ballistic Missile Center, Jan-Jun 59, I, pp I, 48-52.

23. Chief of Staff's Policy Book, 1960, USAF and Related Space Activities, Item 129, Tab E.

24. ARDC Presentation of a Plan for an Advanced, Integrated Space Study Program, 30 Jun 58. Three of the study series had not been released

at this time, but they were sufficiently far advanced for ARDC representatives to speak of them with a knowledge of their content.

25. Statement by Boushey before McCormack Cmte, p 525; SAC brochure, 13 Aug 59, subj: SAC Space Concepts.

26. Memo, Col E.K. Kiessling, Hq ARDC to L/Gen S.E. Anderson COMARDC, 17 Nov 58, subj: Space Study Program; note, Maj P.L. Chell, Analysis Div, Hq ARDC to Col E.K. Kiessling, Hq ARDC, 11 Dec 58, subj: Space Study Prog vs NASA. The NASA programs for fiscal years 1959-60 were outlined and attached to Chell's so-called "card."

27. ASSS by B/Gen D.E. Newton, Hq ARDC, 16 Feb 60, subj: Discussion of AF Study Prog with NASA, Tab A, Abstract; Maj Chell, Introduction at AF-NASA Interchange of Info on Lunar Subjs, 16 Jun 60.

28. NASA Release 59-116, 17 Apr 59.

29. Memo, Maj Chell to Col O.J. Poage, Ch/Mission Analysis Off, ARDC, 17 Dec 59.

30. M/R by Chell, 13 May 60, subj: Info Interchange with NASA (SR 183, Lunar Observatory Studies).

31. Ltr Gen T.D. White, C/S USAF to L/Gen R.C. Wilson, DCS/D, 14 Apr 60; ltr, Col N.C. Appold, Sp Asst for ARPA-NASA, Hq ARDC to Col R.D. Curtin, AFBMD, 29 Apr 60.

CHAPTER X

1. Memo, Johnson to COMAFBMD, 10 Sep 58, subj: Redef of WS-117L; ARPA, Space Technology Program Review, 9-15 Sep 59 (hereinafter cited as ARPA Program Review) I, 201-6; Johnson Briefing for JCS, 10 Jan 59, subj: ARPA's Program (hereinafter cited as ARPA Program, 10 Jan 59); Sup to Hq USAF Daily Staff Digest, 6 Mar 59.

2. Ltr, Boushey to COMARDC, 11 Jul 58, subj: Comm & Navigation Satellites.

3. Ltr, Hq ARDC to D/AT, 10 Sep 58, subj: Abbr System Dev Plan 470L; memo, Horner to D/ARPA, 30 Sep 58.

4. Memo, Johnson to Horner, 22 Oct 58, subj: Strat Comm System; ltr, Schriever to DCS/D, 3 Dec 58, subj: Mgt of Comm Satellite Prog; ARPA Program Review, II, 251-53; memo, M/Gen James Ferguson, D/Rqrmts to Col J.E. Kelsey, 8 Feb 59, subj: ARPA Rqrmts Panel Mtg; M/R by Maj H.C. Howard of D/AT, 10 Feb 59, subj: Comm Satellite; memo, Gen C.E. LeMay, VC/S to Deps et al., 6 Mar 59, subj: AF Position on Comm Satellite Prog.

5. M/R by Maj Howard, 17 Dec 58, subj: Comm Satellite; ltr, Swofford to COMARDC, 24 Dec 58, subj: Joint Army-AF Comm Satellite Dev Prog;

memo, Johnson, to COMARDC, 6 Mar 59, subj: Satellite Comm in Polar Regions for SAC; AO 54-59, Amnd 1, 22 May 59.

6. ARPA Program, 10 Jan 59; ARPA Program Review, I, 153-57.

7. Memo, R/Adm J.E. Clark, ARPA to COMARDC, 7 Jan 59, subj: Launching and Data Readout Planning for Cloud Cover Exper; ARPA Program, 10 Jan 59.

8. ARPA Program, 10 Jan 59.

9. Memo, Col R.H. Ellis, Off of D/Plans, DCS/P&P to Dep D/Plans, 20 Jan 59, subj: Man-in-Space Prog.

10. M/R by L/Col B.H. Ferrer, 4 Nov 57, subj: Boost-Glide Concept; DCS/D DD94, 25 Nov 57, subj: Hypersonic Strat Weapon System; M/R by Col Ferrer, Off of D/AT, 6 Oct 58, subj: Dynasoar Mtg with Mr. Horner on 2 Oct 58; memo, Johnson to SOD, 7 Nov 58, subj: Dyna Soar; memo, Col J.R. Finlon, Exec, DCS/D to All Dirs et al., 7 Nov 58, subj: Dyna Soar.

11. ARPA Program, 10 Jan 59; ARPA Program Review, I, 21-87; memo, R/Adm J.E. Clark, Actg D/ARPA to SAF (R&D), 12 Jun 59, subj: Saturn Booster for AF Space Progs.

12. Statement by L/Gen S.E. Anderson, COMARDC, quoted without ref in DOD Staff Study, 5 May 59, subj: Recs against Transfer of Centaur to NASA; ltr, York to Glennan, 18 May 59; AO 19-59, Amnd 5, 30 Jun 59.

13. Ltr, Capt J.G. Lang, USN, Actg DCS/I, NORAD to AFCIN, 4 Jan 58, subj: Space Track.

14. ARPA System Dev Plan for Space Track (a resume), 13 Jan 59.

15. M/R by L/Col W.D. Pritchard, Ch, Test Instrum Div, Hq ARDC, 2 Sep 58, subj: Satellite Tracking & Surveillance.

16. AO 7-58 & 8-58, 20 Jun 58.

17. Memo, L/Gen R.H. Lynn, VCOMADC to Gen E.E. Partride, CINCNORAD, 8 Oct 58, subj: Mgt of Interim Satellite Detection and Tracking System, w/incl, Background Info.

18. Memo, M/Gen H.E. Watson, AFCIN to D/AT, 25 Aug 58, subj: Space Vehicle Surveillance Prog.

19. Memo, Col J.L. Martin of D/AT to NASA, 22 Oct 58, subj: Tracking of (NASA's) Inflatable Satellite; ltr, Johnson to Glennan, 5 Nov 58.

20. Memo, Johnson to COMARDC, 5 Nov 58, subj: Satellite Detection and Surveillance System Dev.

21. M/R by Maj H.C. Howard of D/AT, 21 Nov 58, subj: Space Surveillance; AO 50-59, 19 Dec 58; ARPA System Dev Plan for Space Track, 13 Jan 59.

22. Memo, Boushey to Chmn/Air Def Panel Weapons Bd, 30 Apr 59, subj: Space Surveillance.

23. Statements by L/Gen A.G. Trudeau, Ch, R&D, USA; V/Adm J.T. Hayward, ACNO (R&D); & M/Gen B.A. Schriever, COMAFBMD, in S Hearings before Subcmte on Govt Orgn for Space Activities of Cmte on Aero & Space Sciences, 86th Cong, 1st Sess, Investigation of Governmental Organization for Space Activities, pp 236, 282, 315, & 416-17.

24. Memo, M/Gen H.T. Wheless, D/Plans to C/S USAF, 15 Jan 59, subj: ARPA Progs; memo, L/Gen Wilson, DCS/D to L/Gen J.K. Gerhart, DCS/P&P, 12 Jan 59, subj: AF Mission in Space.

25. S Hearings, Investigation . . . Space Activities, p 137.

26. Ibid., pp 558-59.

27. Hanson Baldwin, "Neglected Factor in the Space Race," NY Times, 17 Jan 60; James Hagerty, Jr, "U.S. Program Has Gone Far in 28 Months," Washington Post, 22 May 60; Clarke Newlon, "We Can Catch the Russians in Space," Missiles & Rockets, 14 Dec 59.

28. NSC 5818, 26 Jan 59, subj: U.S. Policy on Space.

29. NASA, United States and Russian Satellites, Lunar Probes and Space Probes, 1957-1959; AFBMD, Summary of Air Force Ballistic Missile Division Activities in Space, Jun 60.

CHAPTER XI

1. House Rpt 1758, 85th Cong, 2d Sess, The National Space Program.

2. Statement by Gen Putt before McCormack Cmte, p 123.

3. Statement by Gen Schriever before McCormack Cmte, p 648.

4. D/AT, Proposed Military Position on Space, Dec 59.

5. Memo, Horner to D/ARPA, 10 Nov 58, subj: AF Statement of Mil Uses of Space; WSEG Rpt 39, Mil Applications of Artificial Earth Satellites, submitted 13 Jun and considered by JCS 14 Jul 59. The references could be continued indefinitely.

6. SAC brochure, SAC's Space Concepts, 13 Aug 59.

7. Memo, Boushey to Sp Asst to C/S USAF, 6 Dec 57, subj: Missile Hearing, w/incl 1, policy.

8. Putt before McCormack Cmte, pp 100-101.

9. Memo, B/Gen G.W. Martin, Dep D/Plans to Col R.A. Yudkin, War Plans Div, 3 Mar 59, w/incl; memo, M/Gen H.C. Donnelly, Asst DCS/P&P to D/Plans, 21 Mar 58, subj: Prep of a USAF Position on Sovereignty Over Outer Space; Putt before McCormack Cmte, pp 99-136; memo, B/Gen N.F. Parrish, Asst/Coordination to D/Dev Prog et al., 9 May 58, subj: Mil Questionnaire; memo, L/Col J.B. Ramey, Policy Div, D/Plans to D/Ops et al., 12 Jun 58, subj: Policy Paper on Sovereignty Over Outer Space, w/incl; memo, Col S.G. Fisher, Asst Dep D/Plans for Policy to Asst/LR Objectives, 22 Aug 58, subj: Trans of Study on Sovereignty Over Outer Space.

10. Memo, Fisher to Asst/LR Objectives, 22 Aug 58; memo, Col R.S. Abbey, Policy Div to DCS/D et al., 8 Oct 58, subj: Study on Sovereignty Over Outer Space.

11. Memo, CNO to Chmn/JCS, 2 May 59, subj: U.S. Mil Policy on Outer Space.

12. Memo, Chmn/JCS to SOD, 22 May 59, subj: U.S. Mil Policy on Outer Space.

13. Memo, M/Gen R.M. Montgomery, Asst VC/S USAF to M/Gen C.H. Donnelly, Asst DCS/P&P, 16 Dec 59, subj: Draft Statement on AF Policy with Regard to Space.

14. Gen T.D. White, C/S USAF, "Air and Space Indivisible," Air Force, Mar 58.

15. Memo, B/Gen G.W. Martin, Dep D/Plans to L/Gen J.K. Gerhart, DCS/P&P, 13 Jan 59, subj: USAF Respons in Aerospace; ltr, Gerhart to L/Gen Emmet O'Donnell, Jr, DCS/Pers, 15 Jan 59; memo, Col R.E. Richardson, Asst/LR Objectives to M/Gen H.T. Wheless, D/Plans, 25 Nov 58, subj: AF Leadership in Space; USAF in Space, A Policy Statement, 30 Jan 59.

16. Statement by Gen White in H Hearings before Cmte on Science & Astro, 86th Cong, 1st Sess, Missile Development and Space Sciences, pp 73-74.

17. H Hearings before Subcmte of Cmte on Appn, 86th Cong, 1st Sess, DOD Appropriations for 1960, Pt 1, Policy Statements, pp 574-80.

18. Memo, Col R.H. Ellis, Asst Dep D/Plans for WP to Dep D/Plans for Policy, 26 Mar 59, subj: Def of Aerospace; memo, Col S.G. Fisher, Asst Dep D/Plans for Policy to DCS/P&P, 20 Mar 59, same subj.

19. Memo, Wheless to Gerhart, 17 Jan 59, subj: Aerospace Power and National Security, 1960-70.

20. Memo, Boushey to Gerhart, 7 Jan 59, subj: AF Mission in Space; memo, Wheless to DCS/D, 5 Feb 59, subj: AF Objectives in Space.

21. Ltr, LeMay to Power, CINCSAC, 17 Mar 59.

22. Capt C.W. Styer, USN, & Comdr R.F. Freitag, USN, "The Navy in the Space Age," U.S. Naval Institute Proceedings, Mar 60.

23. Boyd Hill, "The Inglewood Complex: Brawn and Heart of the AF Ballistic Missiles," AMC Worldwide, Apr 59; Hist, 1st Missile Div, Jan-Jun 59, Vol I.

24. Styer & Freitag.

25. USN-USAF Agreement for Coordinated Peacetime Operation of the Pacific Missile Range, 5 Mar 58.

26. Hist, 1st Missile Div, I, 126-52.

27. Ibid.

28. Ltr, Col G.S. Curtis, Senior AF Rep at PMR to Col G.S. Brown, Exec Off, 1st Missile Div, 13 May 59.

29. Ltr, Power to White, 13 Dec 58.

30. Signed Statement by Adm Arleigh A. Burke, CNO, 29 Jan 59, inserted in H Hearings, DOD Appropriations for 1960, Pt 1, Policy Statements, pp 736-39.

31. Hist, 1st Missile Div, I, 126-52.

32. Ibid., pp 120-29, 138-40.

33. Memo, B/Gen J.F. Whisenand, USAF, to Chmn/JCS, 16 Feb 59, subj: Priorities for BM & Sat Progs, w/App, draft memo, SOD to Pres, same subj; memo, Donald A. Quarles, Actg SOD to NASC, 26 Mar 59, subj: Priorities for Sat Progs; Note by the Secys, Action on JCS 2012/135, J-5 2012/135/1, 19 Mar 59, subj: WS-117L Infrared Sat Prog; memo, Martin, Actg D/Plans to C/S USAF, 15 Apr 59, subj: Priorities for BM & Sat Prog; memo, JCS to SOD, 12 May 59, subj: Priorities for Space Progs; memo, Wheless to C/S USAF, 12 May 59, same subj; memo, McElroy to Secys/Army, Navy, AF, 27 Apr 59, subj: Top Nat Priorities Progs.

34. ASSS, Col R.H. Ellis, Asst Dep D/Plans for WP to DCS/P&P, 17 Apr 59, subj: Status of WS-117L; memo, Donnelly to VC/S, 17 Apr 59, subj: Status of WS-117L.

35. Ltrs, Schriever to Gerhart & Wilson, 18 May 59.

36. Memo, Col Butcher to Dep D/Plans, 11 Jun 59, subj: Coord of Sat & Space Vehicle Ops; ASSS, Wheless to DCS/D, 9 Jun 59, subj: Reply to Gen Schriever.

37. Memo, Adm Burke to JCS, 22 Apr 59, subj: Coord of Sat & Space Vehicle Ops, w/Encl, draft memo, JCS to SOD, same subj.

Notes to pages 199-203.

38. Memo, C/S USAF to JCS, 12 May 59, subj: Coord of Sat & Space Vehicle Ops; memo, Wheless to C/S USAF, 14 May 59, subj: Navy Proposal for a Single Mil Agency for Coord of Sat & Space Vehicle Ops.

39. Ltr, Glennan to McElroy, 25 May 59.

40. Memo, Boushey to Swofford, 1 Jun 59, subj: Global Tracking; Col Otto Haney, Off D/Dev Plang, 4 Jun 59, subj: Comments on Ltr of Dr. Glennan to SOD.

41. M/R by Col John Martin, 17 Jul 59, subj: Proposed Orgn for Support of Project Mercury.

42. Draft DOD Dir, 24 Jul 59, subj: Asgmt of Respons for DOD Support of Project Mercury.

43. Memos, McElroy to JCS, 29 May 59, subjs: Asgmt of Operational Respons for an Interim Sat Early Warning System; Asgmt of Respons for an Interim Sat Navigation System; Asgmt of Operational Respons for Phase I of a Sat Recon System; Asgmt of Operational Respons for an Interim Sat Detection System. See also memo, JCS to SOD, 24 Jul 59, subj: Coord of Sat & Space Vehicle Ops; M/R by B/Gen G.W. Martin, 13 Aug 59.

44. Memo, McElroy to JCS, 18 Sep 59, subj: Coord of Sat & Space Vehicle Ops.

45. H Rpt 1758, 85th Cong, p 8.

46. Statement by Wilson in H Hearings before Cmte on Science & Astro, 86th Cong, 2d Sess, Review of the Space Program, p 481.

47. Statements by Fulton & Merchant in Review of the Space Program, p 11.

GLOSSARY

A	Administration
ABMA	Army Ballistic Missile Agency
ACNO	Assistant Chief of Naval Operations
Advsy	Advisory
AEC	Atomic Energy Commission
Aero	Aeronautics
AFBMD	Air Force Ballistic Missile Division
AFCRC	Air Force Cambridge Research Center
AFFTC	Air Force Flight Test Center
AFMDC	Air Force Missile Development Center
AFOSR	Air Force Office of Scientific Research
AFPC	Armed Forces Policy Council
AMC	Air Materiel Command
Anx	Annex
AO	ARPA Order
AOMC	Army Ordnance Missile Command
ARDC	Air Research and Development Command
ARPA	Advanced Research Projects Agency
ARS	Advanced Reconnaissance System
ASAF	Assistant Secretary of the Air Force
ASOD	Assistant Secretary of Defense
ASSS	Air Staff Summary Sheet
Astro	Astronautics
Balwards	Ballistic Weapons and Development Supporting System
BM	Ballistic Missiles
BMEWS	Ballistic Missile Early Warning System
BMTS	Ballistic Missile Test System
Bomi	Bomber Missile
Brats	Ballistic Research and Test System
BRL	Ballistic Research Laboratory
BuAer	Bureau of Aeronautics (Navy)
CEFSR	Committee for Evaluating the Feasibility of Space Rocketry
CGM	Committee on Guided Missiles
C-MLC	Civilian-Military Liaison Committee
D/AT	Directorate of Advanced Technology
DD	Development Directive
DDR&E	Director of Defense Research and Engineering
Def	Definition
fps	feet per second
GM	Guided Missiles
GOR	General Operational Requirement
HETS	Hypersonic Environment Test System
IDA	Institute for Defense Analysis
IGY	International Geophysical Year
IOC	Initial Operational Capability
JPL	Jet Propulsion Laboratory
JRDB	Joint Research and Development Board
JSPC	Joint Strategic Plans Committee

MIS	man-in-space
Mouse	Minimum Orbital Unmanned Satellite of Earth
M/R	Memo for Record
NACA	National Advisory Committee for Aeronautics
NASA	National Aeronautics and Space Administration
NASC	National Aeronautics and Space Council
NIE	National Intelligence Estimate
NORAD	North American Air Defense Command
NOTS	Naval Ordnance Test Station
NRL	Naval Research Laboratory
NSC	National Security Council
OCB	Operations Coordinating Board
ODM	Office of Defense Mobilization
OSAF	Office, Secretary of the Air Force
OSD	Office, Secretary of Defense
Plang	Planning
PMR	Pacific Missile Range
P&O	Plans and Operations
P&P	Plans and Programs
Prog	Program; Programming
RDB	Research and Development Board
Robo	Rocket Bomber
SAB	Scientific Advisory Board
Sat	Satellite
SOD	Secretary of Defense
SR	Study Requirement
TCP	Technological Capabilities Panel
TEG	Technical Evaluation Group
Tiros	Television Infra-Red Observing Station
TN	Thermonuclear
WADC	Wright Air Development Center
WADD	Wright Air Development Division
WDD	Western Development Division
WP	War Plans
WSEG	Weapons Systems Evaluation Group

DISTRIBUTION

HQ USAF

1. SAF-OS
2. SAF-US
3. SAF-RD
4. SAF-AA
5. SAF-LL
6-7. SAF-OI-1
8. SAF-SS
9. AFAAC
10. AFADA
11. AFBSA
12. AFCOA
13. AFCVC
14. AFDAS
15. AFESS
16. AFIIS
17. AFJAC
18. AFMSG
19. AFNIN
20. AFOAP
21. AFODC
22. AFORQ
23. AFPDC
24. AFRDC
25. AFRDP
26. AFRDD
27. AFRRP
28. AFRST
29. AFSDC
30. AFSPD
31. AFXDC
32. AFXOP
33. AFXPD

MAJOR COMMANDS

34-35. ADC
36. AFCS
37. AFLC
38-44. AFSC
45. ATC
46. AU
47. AFAFC
48. AAC
49. USAFSO
50. CONAC
51. HEDCOM
52. MATS

53. OAR
54. PACAF
55-57. SAC
58-60. TAC
61. USAFA
62-63. USAFE
64. USAFSS

OTHER

65-66. ASI
67-71. ASI (HAF)
72-76. ASI (HA)
77-100. AFCHO (Stock)

DISTRIBUTION

HQ USAF

1. SAF-OS
2. SAF-US
3. SAF-RD
4. SAF-AA
5. SAF-LL
6-7. SAF-OI-1
8. SAF-SS
9. AFAAC
10. AFADA
11. AFBSA
12. AFCOA
13. AFCVC
14. AFDAS
15. AFESS
16. AFIIS
17. AFJAC
18. AFMSG
19. AFNIN
20. AFOAP
21. AFODC
22. AFORQ
23. AFPDC
24. AFRDC
25. AFRDP
26. AFRDD
27. AFRRP
28. AFRST
29. AFSDC
30. AFSPD
31. AFXDC
32. AFXOP
33. AFXPD

MAJOR COMMANDS

34-35. ADC
36. AFCS
37. AFLC
38-44. AFSC
45. ATC
46. AU
47. AFAFC
48. AAC
49. USAFSO
50. CONAC
51. HEDCOM
52. MATS
53. OAR
54. PACAF
55-57. SAC
58-60. TAC
61. USAFA
62-63. USAFE
64. USAFSS

OTHER

65-66. ASI
67-71. ASI (HAF)
72-76. ASI (HA)
77-100. AFCHO (Stock)

www.ingramcontent.com/pod-product-compliance
Lightning Source LLC
Chambersburg PA
CBHW082117230426
43671CB00015B/2719